MW01156553

Sustainability
Management
Handbook

Sustainability Management Handbook

Shirley J. Hansen, Ph.D.
James W. Brown, P.E.

THE FAIRMONT PRESS, INC.

CRC Press
Taylor & Francis Group

Library of Congress Cataloging-in-Publication Data

Sustainability management handbook / [edited by] Shirley J. Hansen, James W. Brown.
   p. cm.
Includes bibliographical references and index.
ISBN-10: 0-88173-644-9 (alk. paper)
ISBN-10: 0-88173-645-7 (electronic)
ISBN-13: 978-1-4398-5195-1 (Taylor & Francis : alk. paper)
    1. Sustainable engineering. 2. Project management. 3. Manufacturing processes--
Environmental aspects. I. Hansen, Shirley J., 1928- II. Brown, James W., 1950-

TA190.S895 2011
620--dc22

                                   2010038783

Sustainability management handbook / [edited by] Shirley J. Hansen, James W. Brown.
©2011 by The Fairmont Press, Inc.. All rights reserved. No part of this publication may
be reproduced or transmitted in any form or by any means, electronic or mechanical,
including photocopy, recording, or any information storage and retrieval system, without
permission in writing from the publisher.

Published by The Fairmont Press, Inc.
700 Indian Trail
Lilburn, GA 30047
tel: 770-925-9388; fax: 770-381-9865
http://www.fairmontpress.com

Distributed by Taylor & Francis Ltd.
6000 Broken Sound Parkway NW, Suite 300
Boca Raton, FL 33487, USA
E-mail: orders@crcpress.com

Distributed by Taylor & Francis Ltd.
23-25 Blades Court
Deodar Road
London SW15 2NU, UK
E-mail: uk.tandf@thomsonpublishingservices.co.uk

Printed in the United States of America
10 9 8 7 6 5 4 3 2 1

10: 0-88173-644-9 (The Fairmont Press, Inc.)
13: 978-1-4398-5195-1 (Taylor & Francis Ltd.)

While every effort is made to provide dependable information, the publisher,
authors, and editors cannot be held responsible for any errors or omissions.

Contents

Foreword

Sustainability can deliver great value to our environment, economy and our way of life. The question is: In 2030 will all these benefits be a part of the way we live and do business, or will "sustainability" be remembered as a buzz word from back in the beginning of this century?

Facilities and processes are not impervious to passing fads. Already we have surveys that show growing numbers are tired of "green." *Greenwash* is taking its toll. The challenge will be to *sustain* "sustainability. "

Managing sustainability effectively and demonstrating positive results are the best ways to convince people that the sustainability approach is sound and not a passing fancy. This means we need a solid consensus of what we are talking about. A broadly accepted working definition of sustainability is at the core. It's incredibly difficult to manage what seems so nebulous. When asked to define sustainability, too many confuse the *why* with the *what.*

Discussions of sustainability too often become a finger pointing exercise, as we attempt to identify all the things we do wrong. Finger pointers cite the loss of wetlands, deforestation, the greenhouse effect, and global warming as well as air and water pollution. Unfortunately, action at the local level is frequently viewed as limited and the global advocates see the answer as sweeping government intervention, mandates, restraints and taxes.

Sustainability can, and should, have a more positive approach. The very presumption, which declares that opportunities at the local level are limited, fosters the attitude that it is someone else's problem. Actions at the local level include what we do as individuals, organization, communities and corporations. It is sustainability that is doable and *MANAGEABLE*—by us. That is the focus of this book.

In writing Chapters 1, 3 and 4, I have drawn from over 30 years in the business. The material is basically a discussion with the reader. As such, I have chosen to step over the line and write much of it in the first and second person. After being in the business so long, I've formed a number of opinions and do not hesitate to share a few I feel strongly about.

My co-author, Jim Brown, has chosen to share some of his experiences in the first person as well, so the authors are identified at the head of each chapter. That way you know who the "I" is in each case.

Sustainability is so much more than technology. It has a huge human component—attitude, behavior and practices. All of which demand effective leadership. Managers of sustainability need to be MANAGERS first. To do so, they must have a broad spectrum of knowledge and a respected seat at the management table. Sustainability manager is not a new name for energy managers.

It is gratifying that some major corporations have designated chief sustainability officers, or vice-presidents of sustainability, but that is not enough. The concept needs to permeate industry. This book is dedicated to providing the knowledge and the tools to manage sustainability effectively.

To that end, I am extremely pleased with the caliber of people who have agreed to share their experience and expertise. A brief bio and contact information for each contributing author appears in Appendix A.

To help establish what sustainability is, Robert J. Dixon offers a world perspective in Chapter 2. Bob is the Sr. VP and Global Head of Efficiency and Sustainability for Building Automation, Siemens, with responsibilities that take him to the far corners of the Earth. Long before others noticed, Bob was citing environmental concerns as a major driver for energy efficiency and energy service industry growth.

A command of sustainability requires clear well thought out policies. Dr. Stephen Roosa, in his book, *Sustainability Development*, answered this call. In Chapter 5, Steve has generously shared his thoughts on sustainable facilities, which have been adapted from that book.

Risks are always with us and seem even more threatening as we venture into the uncharted waters of sustainability, so insurance becomes a critical component in managing an area rife with many emerging technologies and innovations. I am particularly pleased that Bob Sansone has drawn on his wide-ranging experience in the risk analysis and insurance field to offer guidance regarding green insurance.

Being on the cutting edge in a field, such as sustainability, always poses the question, "How are we doing?" Fortunately, the U.S. Green Buildings Council has given us the LEED yardstick to measure our progress. Nick Stecky, certified LEED specialist and recognized expert in his field, shares some valuable insights on making LEED an effective tool for chief sustainability officers (CSOs).

This foreword would not be complete without mentioning how very fortunate I am to have James W. Brown as a co-author. As the head of Energy Systems Associates, Jim is regularly in the trenches with his team, and gives much needed practicality to the contents of this book. An increasing portion of ESA's business is retro-commissioning, so Chapter 7 provides especially useful guidance in an increasingly important part of a sustainability program. Jim is an exceptional engineer, who has repeatedly kept his firm a step ahead of the profession. He taught auditing when it was in its infancy and was one of the first to acknowledge that while traditional auditing had served us well, it was no longer good enough. This led to our co-authoring *Investment Grade Energy Audits: Making Smart Energy Decisions.* Early on, Jim recognized that a single audit report or a given design implementation does not give clients the continuing guidance they need, so ESA pioneered the use of master planning. Authoring the chapters on auditing, commissioning, and master planning, Jim has brought this wealth of experience and expertise to bear on three fundamental aspects of sustainability management

The authors realize that page after page of text can be rather off-putting, and breaking up the print makes the whole thing less daunting. We also appreciate that nothing makes a point better than a bit of humor. Both of these concerns have been answered by the artistic endeavors of Stephen Hansen. I have been asked how one gets the work of such a talented artist, who has had work in the Smithsonian and in private collections around the world, to provide his work in a technical book. The answer? It really helps if you are the mother.

Books don't just happen. A lot of "blood, sweat, tears," and *editing*, go into such a work. Once again, my business colleague and partner in life has offered his magic touch. For so very many things, including the editing of this book, I am deeply indebted to James C. Hansen.

The need for managers to be informed has been stressed in the early chapters. Clearly, this must include some discussion of the critical topics of the day, such as global warming and climate change. Any opinions or perceived bias expressed are strictly my own and do not necessarily reflect the thoughts of the contributing authors, that of my co-author, or any of their respective firms.

As you move ahead managing your sustainability programs, it is hoped that the sustenance provided within these covers will **sustain** your sustainability efforts.

Shirley J. Hansen

Chapter 1

Bringing Sustainability Down To Earth

Shirley J. Hansen

Since the beginning of recorded history, man has pretty well taken from our Earth whatever he desired. We have become a voracious "consumption" species with little or no regard for the consequences. In large measure, we have acted like over-indulged teenagers. Too often, our "progress" seems to be measured by possessions. We wrap ourselves in denial and refuse to truly look around.

Unfortunately, many alarmists have played right into that denial by exaggerating the situation, feeding us unfounded data, and making claims not backed by solid scientific data. Rather than helping their cause, they have hurt it. Perhaps irreparably.

Problems associated with shaky data have been compounded by the zealots who live in a world of absolutes. Declarations abound: "There shall not be another coal generating power plant." "There will be no off shore drilling." "We must tear down all dams." "No living creature shall become extinct. " Sustainability and our Earth need us to be vigilant, not vigilantes.

There is growing awareness that we must do a better job of taking care of the world around us. We are beginning to wake up to what we might be dumping on our children, grandchildren and those beyond. Our insatiable appetite to consume all Earth offers is rapidly depriving future generations of the choices that should be theirs to make. Some of us are also beginning to realize that we cannot turn it around overnight. There is a wonderful old Chinese proverb that tells us if we want to move a mountain, we should start with a few stones at a time.

If we are to start with a few stones, our efforts to sustain our world will require careful management. Which stones do we move first? How far can we move them? Who will we ask to help us move those stones? How much will it cost?

Managing sustainability could very well be the greatest challenge of our lifetime, of any lifetime. We are asking people to seriously consider the way they live—and change it.

Sustainability *management* requires a major focus on the key role people play in making sustainability happen. Changing the behavior of many demands strong leadership. That leadership needs to be backed by a solid knowledge base, the ability to make key judgment calls, effective communication skills, and a command of issues vital to the organization and to a sustainable world.

In the process, we will need to puzzle over some hard questions. Can we go to the developing countries and tell them they should not want what the developed nations have? How do we

THE PROBLEMATIC PIECE
OF THE PUZZLE

explain to loggers in Washington State that they lost their liveli-hood to save the spotted owl, which supposedly lived only in old growth fir trees? Especially when it turned out that their habitat is spread far and wide? When confronted with this information, the spotted owl instigator's comment was, "We wondered when somebody would finally notice." People like that really hurt the sustainability movement. Particularly when the folks in the "Less-er Washington" refuse to admit they made a mistake!

When it comes to preserving the species, there is an enclave within the sustainability movement that rather vociferously de-clares that nothing shall change—unless they want it to. The prob-lem is that change is one of the most consistent factors of life. And we have little control over much of that change. Ice Ages seem to just come and go. The extinction of a species has always been part of the natural order.

Some of our wishes fall into the old category of being careful what you wish for. We all love those beautiful white polar bears, even those drinking Coca-Cola. We worry that their numbers may be declining—even when some studies show they are not. But few of us would really want a sloth as big as an elephant in our cherry tree, dinosaurs roaming around our backyard, or pterodactyls hov-ering overhead ready to snatch up our adorable pet cocker spaniel.

Where do we draw the line? How do we achieve balance? How do we manage sustainability ?

It should not be done by fear and dire predictions. It should not be done by exaggerating future sea levels. Or, gazing into a crystal ball to foretell how many millions of lives will be lost while shorelines crumble. It should not be done in moments of hysteria by government edicts, mandates or unwarranted intervention. In fact, if it gets to that, it means we have not effectively addressed those concerns in a timely manner.

To really be "sustained" over time, sustainability requires quality leadership growing out of true science and a rational, bal-anced approach.

The folks in East Anglia may have been trying to do good, but unfortunately they have done serious harm. They once more re-fueled the "Global Warming" debate, and much of the good information they may have had has lost credibility in the scientific community. We have expended a great amount of energy on the "Global Warming" debate. Energy that could have been better spent making strides in areas that could reduce those actions of man which may be causing climate change.

We can all agree that the Earth experiences climate change and has for millions of years. We all know that some serious degradation is taking place; degradation that needs immediate attention. Valiant attempts are being made to determine to what extent these problems are man-made. That information might prove useful; however, it also could be using time and money that could better be spent solving problems that demand immediate attention.

Everyone recognizes that environmental problems exist that are detrimental to our health. The statistics are there. It is a known fact that the people, who live around the coal mines in Krakow, have a life expectancy that is eight years shorter than those who live in other parts of Poland. Sadly, the people there know it also, but point out they have to make a living.

There are enough "Krakows" to keep us busy for a long time. At this point in time, we need to gather our intelligence, prioritize our concerns, and get to work on what we can do.

Space exploration provided us incredibly beautiful pictures of our planet. It also drove home the message that our planet is relatively small, precious to us and vulnerable to our actions. No attempt is made in this book to catalog all the problems we face, nor is an attempt made to cite statistics and sources. When you picked up this book, you already knew that we have a problem, so the critical question before us is what are we going to do about it.

A FEW STONES AT A TIME

We need to worry globally and act locally. Effective management can only manage what it can get its arms around. Administrative theory has a wonderful expression, "span of control." If we are to make a difference, we need to recognize our limits, our *span of control*. We can move mountains, but only if we each grab a few stones and get to work.

The questions of the day are: How do we get organizations, particularly businesses, to embrace sustainability as a business strategy while maintaining a competitive advantage? How do we establish senior leadership and create the programs designed to reduce carbon footprints with the sense of urgency our Earth needs?

Understandably, advocates are getting impatient and feel a sense of urgency to change things more radically and rapidly. There is a sense of panic among those who sense we are running out of time—that if we don't take actions today, serious harm will be inflicted on our planet. Harm, which may take decades to remedy, or even cause irrevocable damage. Weighing the options, gauging the best use of limited resources, and prioritizing contemplated actions are all critical management responsibilities. These challenges are the underlying reason for this book.

When "energy" emerged in the 1970s as a major concern, "energy experts" climbed out of the woodwork equipped with the new jargon of "retrofit" and "ECM." Unfortunately, few of the "experts" had the substantive knowledge to back it up. Similarly, today we have the green charlatans lurking in the bushes while they spout greenwash and make unfounded green claims. If we are to believe them, green deposit slips makes a bank eco-friendly. Fashion houses offer eco-chic—whatever that means. Then, there's the assertion that plug-in cars do not use energy! Makes one wonder where the plug-in "juice" comes from.

Starved for energy efficiency technology information in the 70s, the energy managers quickly became regarded as techies and were assigned to the boiler room. Most have yet to crawl out. Unless we clearly delineate our goals and arrive at a working definition, managers of sustainability are apt to suffer a similar fate.

Frequently in the literature, "sustainability" is used interchangeably with renewable energy. Renewable energy is certainly important to our sustainability efforts. In fact, energy efficiency is even more vital to those efforts. But sustainability is much broader than any energy concerns. Sustainability cuts across all resources and dictates that we use them wisely. It's doing the very best we can with what we've got, and staying mindful of the consequences of our actions.

There is no question that energy is a vital component of any sustainability plan. Our economies are absolutely dependent on energy and to the extent that we rely on fossil fuels, emissions are a key concern. This brings into focus the crucial element of alternative fuels and renewable energy. But worries about water further underscore the fact that sustainability is not a one horse race. Efforts to maintain our planet for future generations are multi-faceted and complex.

The U.S. Environmental Protection Agency (EPA) captured the essence of sustainability by defining it as "…policies and strategies that meet society's present needs without compromising the ability of future generations to meet their own needs."

In discussing how the term has evolved over the last 30 years, EPA also offers a public policy perspective as "… the satisfaction of basic economic, social, and security needs now and in the future without undermining the natural resource base and environmental quality on which life depends." The Agency also states that the goal of sustainability from a business perspective is to "… increase long-term shareholder and social value, while decreasing industry's use of materials and reducing negative impacts on the environment."

One trusts that the EPA meant the *relative* use of materials, for growing industries are expected to use more materials. For a regulatory agency, these definitions are quite generous as they leave a lot of room to define "needs" from the user perspective. Nevertheless, it places a daunting task before us. Thoughts on how to implement these goals is the focus of Chapter 3 and, of course, addressed throughout the book.

THE "GREEN" GAUGE

If we are to effectively determine how we are doing, we need some benchmarks. It has been a long time coming but a sustainability standard for manufacturers was finally unveiled in August 2010. Put forth by UL Environment and GreenBiz.com, ULE 880 is considered a living document. The authors are now soliciting comments.

In its preliminary stage, the standard has five domains:

- **Sustainability Governance**: how an organization leads and manages itself in relation to its stakeholders, including its employees, investors, regulatory authorities, customers, and the communities in which it operates

- **Environment**: an organization's environmental footprint across its policies, operations, products and services, including its resource use and emissions

- **Workplace**: issues related to employee working conditions, organization culture, and effectiveness

- **Customers and Suppliers**: issues related to an organization's policies and practices on product safety, quality, pricing, and marketing as well as its supply chain policies and practices

- **Social and Community Engagement**: an organization's impact on its community in the areas of social equity, ethical conduct, and human rights.[1-1]

The ULE 880 has 102 questions, or indicators, that are broken out into 18 for Governance, 45 in Environment, 15 in Workforce, 15 in Customers and Suppliers, and 9 in Social and Community Engagement. The number of questions does not indicate the weight each category holds. The weighting is currently assigned at 80 points for Environment, 40 each for Governance and Customer/ Suppliers, and 20 each for Workplace and Social/Community. In addition, there are 18 "Innovation Points" with 3 points each for six different indicators that are designed to reward companies for going above and beyond the standard.[1-2]

According to its lead author, Ms. Rory Bakke, director of sustainability at Greener World Media, "The goal of ULE 880 is to level the playing field for companies, as well as their customers, investors, employees and others, in assessing what it means to be a sustainable business."[1-3]

While aimed at manufacturers, the standard is expected to influence the sustainability movement across all sectors.

SUSTAINING WHAT?

As we view the global implications of managing sustainability, we inevitably come to the question: Should everything be sustained? Do we want cigarettes to prevail? Should we allow killer bees to spread unhampered? Do we protect malaria? Or the mosquito?

With such blatant examples, the answers are obvious. However, we can quickly go from the obvious to grey areas. If we decide everything should NOT be sustained, where do we draw the line? What criteria do we use? And who gets to decide? Once you are recognized as the chief sustainability officer in your organization, someone will pose such questions—as ridiculous as some might seem. And as difficult as it might seem, you will be expected to have some answers.

In the natural order of things, Earth has been evolving for millions of years. It is incredibly egocentric of us to think that because we are here and have become more environmentally conscious, we should, or could, bring everything to a standstill. Extreme measures that assure no species will become extinct could be the equivalent of that standstill. It is not natural to stop evolution. Mankind, the good and the bad, is part of that evolution.

When we look around, sustainability should be about preserving what is—including evolution; not bringing things to a dead stop.

Sustainability management demands that we halt, or at least diminish, actions that damage or destroy our environment. But what are the consequences of stopping one action? What are the implications for other actions? What are the trade-offs? What are the costs?

There is much we don't know about preserving our environment. Too often, we rush in only to find out later that the actions were not well thought through. At one time, wood burning stoves were considered the answer to many environmental concerns. For the "back to nature" crowd, it was the right thing to do. Now, we are well aware of the pollutants emitted from wood burning. Wood is a renewable fuel source, but it is *very slow* to renew itself. It also reminds us that renewable are not necessarily low-emitters.

To save a small endangered fish in California, should we have sacrificed acres upon acres of valuable farm land? In letting a large region go arid, have we sustained all the flora and fauna that would have flourished there? Did we even take time to consider that, in saving the fish, the lack of water could have made some critical flora extinct? What have been the social consequences of the lost agricultural businesses and products? What are the economic consequences to the farmers who have worked the land for generations? Or, those whose living depends on work in the fields or in nearby processing plants?

These examples illustrate the "rush to solution" problems that absolutes and government intervention create. In the U.S., the Delaney Amendment, which protects living creatures from extinction, is a good law on the face of it, but ignores consequences. It offers little in terms of rationally weighing potential impact and alternatives.

Government intervention is rife with unintended consequences. Government action to resolve those consequences is usually another law—with its own unintended consequences. Many years ago, Dutch Cleanser had a little Dutch girl holding a can of Dutch Cleanser, which had a smaller Dutch girl holding a can of Dutch Cleanser and on—and on. Government intervention is all too much like the little Dutch girl—with no end in sight.

MAKING CHOICES

Every time we push down the toaster lever, start the lawn-mower, turn up the thermostat, or press the accelerator, we require energy. Over 80 percent of that energy is comes from the ground.

We can get that energy from a far off land where pollution is indiscriminant and punishment to our planet is often irreparable. Or, we can get it out under our regulations and our supervision in an environmentally conscious way.

When the U.S. insists on buying our fuel from other countries, it gives away control over its economy, weakens its national security, gives away jobs, and seriously jeopardizes its energy independence. We can use our work force and create jobs. Or, we can ship off billions of dollars every year, so other countries can have those jobs as they profit from our consumption.

An alternative is aggressively developing renewable fuels domestically. An attractive option, but it won't happen overnight. We have many years ahead where we must depend on energy from the ground. The choice before us today is not oil vs. renewables; *it is OUR oil vs. THEIR oil.* In the rush to condemn off shore oil drilling after the serious damage caused by the British Petroleum (BP) explosion in the Gulf of Mexico, a few pertinent facts were ignored. For example, more spillage in the oceans come from tankers than from wells. Less off-shore drilling means more tankers.

On every front, we see the desperate need for strong management and qualified leadership. Knee-jerk reactions need to be replaced with carefully thought out alternatives.

We'd all like to see home grown versions of environmentally attractive energy sources. Hopefully, we will ultimately get there. But we have a long ways to go. Renewables cannot meet our needs today. Or even tomorrow. It will take time, ingenuity and resources. In the meantime, our economy keeps demanding more energy and we must have fossil fuels to meet that demand. When we try to kid ourselves that we can do it now with renewables, we just give more power to those who supply the fuels. And many of those suppliers are not our friends. They are in a position to do us, and our planet, great harm. As long as we are in denial, we are furthering their cause.

Similarly, every time we turn on the faucet, we expect water to come out. Today, most of us take water for granted much as we did energy in 1970. A portion of Chapter 3 is devoted to water and wastewater management, because a global "water crisis" is imminent. It is not unusual today to hear those, who live in areas fed by the Colorado River, voice resentment about those folks in California getting the Colorado's water. A looming and daunting responsibility facing many chief sustainability officers will be the wise use of water resources.

KEEPING AN EYE ON "THEM"

There are many actors in the sustainability movement. Not all are equally effective. The actions of some in the far corners of the world, however, can affect your life—good and bad. Just as the Delaney Amendment has drastically changed life for farmers in California or loggers in the Pacific Northwest, the push and incentives from Washington, D.C., for corn-based ethanol wreaked havoc with communities in the Midwest, which suddenly found themselves seriously concerned about water resources. Serious enough to take it to court.

The domino effect became painfully obvious when the knee-jerk moratorium on drilling in the Gulf put thousands out of work. Then, to put a band-aid on the hemorrhaging of jobs due to the moratorium, the U.S. government pressured BP to set aside a very large sum of money to pay the laid-off workers. Little, or no, thought was given to the impact on the retirement income in the UK, which is heavily supported by BP. Whether prompted by politics or a genuine fear of contamination from oil, the action has had far reaching effects. One domino topples another.

On another front, U.S. and Britain, which are the two largest members of the World Bank, have been vocal over loaning South Africa $3.75 billion to build a coal-fired plant. ESKOM, the

South African state-run utility, has proposed developing a plant with the best supercritical "clean coal" and carbon storage technology available. The U.S. has argued for a "no or low carbon" energy option. Such pressure caused representatives of South Africa, China and India to jointly write the World Bank president, Robert Zoellick, protesting the "unhealthy submissiveness of the decision-making process in the bank to the dictates of one member country." The loan to South Africa has been made, but the contentious nature surrounding it illustrates the mixed needs and issues facing us.

Many viewed the Copenhagen Conference as a disappointment as the negotiations failed to deliver little more than a statement of intent. Some were even more disappointed because no specific emissions reduction targets were set. At the same time, the implications for the utility industry and clean energy technology were quite significant.

There is a sense that private investment will ultimately drive the transition to a lower carbon economy. Experts expect alternatives to cap and trade will emerge with investments and low-carbon deployment showing some growth in the second decade of this century. A large global carbon market is not expected near term, nor do the soothsayers expect any market changing carbon taxes to materialize soon.

As long as cap and trade legislation, or any climate bill, looms on the horizon, a sense of uncertainty in the utility industry and the broader energy industry will prevail. As Lewis Hay, chief executive of power for the FPL Group, Inc., observed that utilities need to know what climate change legislation will look like before any significant investment in nuclear or renewable power is made. From another perspective, Tom Albanese of Rio Tinto, which does long-term planning for its mining as much as 30 years ahead, declared major investments require some assurance the market is going to be there.

Political uncertainties cause an investment paralysis across the industry. That paralysis echoes down to the day-to-day choices the chief sustainability officers must make.

EFFECTIVE LEADERSHIP

Surrounded by uncertainties and unknowns, the future of sustainability desperately needs effective leadership. We cannot afford waves of hysteria or a grasping at straws to weave a weak fabric that will not hold up over time.

Too often, sustainability advocates focus on a single problem—and a single solution. It is not that simple. Sustainability does not happen in a vacuum. Everything is connected. Nothing happens without affecting something else.

Sustainability management cannot focus on one idea, one technology or one energy source. Management requires *informed* decisions. "Informed" decisions require a solid knowledge base. To effectively lead, management must know the options available as well as the connections and consequences of those actions. We definitely need the knowledge and ability to analyze how certain actions will impact the world around us—even when the impact may be far away. We certainly don't need another "ethanol" taking food off the table in Mexico.

If we are to bring sustainability down to Earth, a practical approach that considers the broader ramifications of our decisions is needed.

Quality management has several key components, including command of a broad range of topics plus the communications skills needed to share that knowledge with a broad range of people.

Fundamental to making all this happen is effective leadership. Unfortunately, only a small part of truly effective leadership can be taught. Most of it is innate. In selecting a person to manage sustainability, there is a broad continuum ranging from people with great knowledge of sustainability issues and scant leadership skills to people who inherently lead effectively but are not familiar with the topics involved. An organization that wants a top quality sustainability program needs to seek leadership qualities. The topics can be taught.

Many peripheral issues to sustainability, such as global warming, have become controversial. At play are not only the sources of energy, but the pipes and wires needed to get energy from the source to the consumer.

The heated discussions surrounding such issues can detract from the fundamental components of sustainability. It can also harden resistance to some needed behavioral changes. Add to that a goodly portion of people who resist change, any change. We can underscore all this even further by noting that when change is foisted upon people too dramatically or too rapidly, it prompts even stronger resistance. Such resistance drastically reduces the likelihood of our achieving the changes we seek. Without effective leadership, this resistance could derail some important sustainability efforts. Some polls already suggest that a large portion of the population is tired of "green." Writing for predominantly "green" readers in the August 2010 *Intelligent Energy Portel*, Energy Insights editor, Paul Mauldin, wrote, "We need more well trained engineers and scientists much more than we need a generation of young people pursuing vague 'green' careers."

To overcome resistance and change behavior, the person responsible for achieving sustainability goals needs to be high enough in an organization to foster change. It is hard to initiate change. It is even harder to initiate change among those who are several levels above you in the organization. The placement on the org chart and/or the title given to the sustainability manager send a very visible message to the rank and file as to how committed top management is to the concept. For maximum effectiveness, the management of sustainability should be on a par with the chief financial officer, the chief information officer and the chief operating officer. Therefore, the term used in this book for the manager of sustainability is the chief sustainability officer (CSO).

When CSOs are pressured by top management, the board or outside groups to take a certain stand, it pays to take the time to see if you really want to be singled out for the "honor." Too often the recognition is to further a certain platform that has little or nothing to do with your current operations or sustainability efforts.

We cannot move far into the world of sustainability without someone raising the issue of energy supply. Important aspects of

A CSO IS BORN

managing our energy needs is discussed in Chapter 3 and our energy supply options in Chapter 4. It will help to first recognize the political environment within which the chief sustainability officer must operate.

POLITICS

A chief sustainability officer (CSO) does not have to be a "died-in-the-wool" politician, but the job does require some political savvy. The CSO must be cognizant of the political world that surrounds and influences much of what can be accomplished.

Whether the CSO is part of a school system, a hospital or a corporation, politics can enable or hinder what we hope to accomplish. Politics and politicians may be a distant concern with limited impact. Or, the CSO might suddenly find him or herself in a maelstrom of activity. In today's political environment, it is more apt to be the latter. Despite all that is going on in Washington, D.C., or your state capital, your organization's internal politics can be the most vital consideration. Situations will vary and times will change, the CSO must be tuned in to the political milieu in which the organization operates.

Energy has become a political tool. The environment has become a political football—often kicked around to serve another political agenda. One of the most telling illustrations of the "football" theory can be found in the criteria used by *Corporate Knights* magazine to select the top Global 100 list of most *sustainable* firms. Quoting directly from GreenBiz, the "performance indicators" were:

- Energy, carbon, water and waste productivity—ratios of sales to total direct and indirect energy consumption, total carbon dioxide and carbon dioxide equivalent emissions, total water use and total was produced, respectively;

- Leadership diversity, which was gauged by the percentage of women on a company's board of directors;

- A comparison of the highest paid executive's compensation to average employee compensation;

- The percentage of total reported tax obligation that was paid in cash;

- A score for sustainability leadership that was based on whether the firm has a sustainability committee and whether a director is part of the group;

- Sustainability remuneration, which was determined by whether at least one senior officer's pay is linked to sustainability;

- Innovation capacity, expressed as a ratio of R&D and sales; and

- Transparency, which was measured by the percentage of data points for which the company provided information and its level of GRI disclosure.[1-4]

You will find no one who is a stronger advocate of women being involved in companies' boards of directors. I also applaud diversity. But no matter how I look at it, I cannot figure out why the presence of women on a company's board of directors should be a key criterion in determining a firm's sustainability level. What prompted this criterion? Who pushed it? What political agenda was involved?

There is an evident bias in some of the other indicators as well. Some are not even very subtle. Yet, the report from *GreenBiz* was very straight forward and no reaction to these criteria was presented.

It is diabolical to consider even the remote possibility that a corporation will seek to put more women on the board, or rush around to pay taxes in cash, to be considered more "sustainable." CSOs, which have their progress measured by *Corporate Knights* magazine, really have their work cut out for them.

It is irrevocable. What "they" do will affect what you can do. Sustainability must move forward within the international political and economic realities. Ironically, we can be restrained by the vulnerability of half-developed regions. Norman Stone in writing *The Atlantic and Its Enemies,* observed that Greece is the pebble announcing the avalanche.

Several astute minds, which keep an eye on such things, have noted that the world seems to be drifting into two camps. This "drift" has been accelerated by the financial crisis, global recession, and may have been weakened by the free-market capitalism case. The coming decade is apt to see the U.S., EU, Japan, Canada and Australia forming a closer alliance to protect themselves against the trend to use state power to freeze out capitalistic ventures. As this evolves into two blocks, they will compete for better political and trade relations with India, Brazil and Mexico. Within this jockeying for position, the need to adhere to our goals of sustainability could easily be lost in the confusion.

Of particular interest in this struggle will be the energy supply situation. When asked what is the biggest oil company, many in the U.S. will respond Exxon Mobil. Not even close! There are 13 state-owned oil companies that have greater reserves than Exxon Mobil. When considered in terms of relative reserves, Exxon Mobil seems pretty small. Saudi Arabian Oil Co. has a 259.9 billion barrel reserve, the National Iranian Oil Company has 136.2 billion barrels, Petroleos de Venezuela 99.4 billion barrels while Exxon Mobil has 7.6 billion barrels. The Saudis reserves are over *32 times* greater than Exxon Mobil. One only has to look at Russia's actions toward Ukraine or comments by Chavez to appreciate that oil is increasingly being used as a political and economic weapon.

There is, however, another side to our evolution that CSOs need to consider. A growing body of thought suggests that man has triumphed by our collective and cumulative intelligence. Anthropologists have found that the tools and artifacts of Neanderthals remain close to their points of origin, while humans that triumphed have left abundant evidence of trade and the merging of ideas. The inventiveness and rate of cultural change of populations appear to be directly linked to the interaction among individuals—and the resulting collective knowledge. If we are to learn from history, neither you nor I can do it alone. Whether we are talking about groups within an organization, a community, a nation, or among nations, it appears strength is in our numbers—and the exchange of ideas. Which further underscores the critical role a CSO's leadership must play in effecting change.

The above thoughts offer a sample of the many factors at play as CSOs strive to make a difference. But it is the real world in which we must operate. We may move only a few stones at a time, but there are days when they will seem like boulders. With a bit of technology and the will, we can even move some huge boulders.

Years ago, I was told that as far as we know life is not a dress rehearsal; it's the real thing. The same can be said of our planet. We are not living an experiment in how to use resources. This is the real thing. As far as we know, we don't get to go around again if we don't get it right this time.

Chapter 2

World Perspective On Sustainability

Bob Dixon

From the beginning of time, mankind has consumed the earth's natural resources. And from the beginning of time until the industrial revolution, the impact of mankind's consumption of the earth's natural resources was minimal. But since the industrial revolution, the impact is becoming more pronounced, and with an ever increasing population, will continue accelerating over time.

Consider the following:

- The average life expectancy in 1950 was 46.6 years, and is expected to increase to 72 years by 2025.

- The world's population is expected to increase from its current level of 6.7 billion to over 9 billion by 2050.

- 50 percent of the world's population lives in cities and will increase to 60 percent by 2030.

- Today 280 million people live in megacities of 10 million residents or more.

- Cities consume 75 percent of the world's energy, are responsible for up to 75 percent of greenhouse gas emissions, and account for 60 percent of world's water use.

- 11 of the 12 years between 1994 and 2005 rank among the 12 warmest since weather observations began.

- Today we face a higher concentration of CO_2 in the atmosphere than in any of the past 350,000 years.

In addition, as developing regions become more industrialized and increase their standard of living, the effect on the consumption of natural resources will be impacted even further.

The scientific community has recognized these trends for some time. The world's governments and policy-making organizations continue to make progress in establishing binding regulatory-based greenhouse gas reduction commitments. Industry has recognized there are significant business opportunities in accepting the challenge of helping the world meet these greenhouse gas reduction commitments, and individuals around the world are changing their personal behaviors to help.

The world has recognized that it is the *"Perfect Storm for Sustainability and Energy Efficiency"* and they are starting to *"Ride the Green Wave."*

THE MARCH TOWARDS SUSTAINABILITY

There are numerous examples of evidence that this march toward sustainability for governments, industry and by individuals is underway. We are past the need to address fundamental infrastructure needs, we are past just focusing on minimizing costs & maximizing efficiencies, and we are marching towards the reduction of greenhouse gas emissions through holistic sustainability.

Government Directives
Government directives can provide guidance and enforcement. The European Union has issued two directives, Directive 2002/91 and Directive 2006/32, specifically targeting the energy performance of buildings, energy end-use efficiency, and energy services. The European Union's National Energy Efficiency Action Plan promotes the improvement of the energy performance of buildings via performance requirements, certification for buildings, and the inspection of heating and cooling systems.

California's Executive Order S-20-04 established the state's priority for energy and resource-efficient high performance buildings. The Executive Order sets a goal of reducing energy use in state owned buildings by 20 percent by 2015 (from a 2003 baseline) and encourages the private commercial sector to set the same goal.

Australia has implemented the 2004 National Framework for Energy Efficiency, 2007 National Framework for Energy Efficiency-Stage 2, and Building Codes Australia-2007.

Government Funding

Government funding can stimulate research and development and implementation. The United States government has funded $3.1 billion for state energy programs, $3.2 billion for energy efficiency and conservation block grants, $5.0 billion for weatherization assistance programs, $300 million for energy efficient appliance rebates, and billions more for smart grid demonstration projects as well as research and development projects.

In the U.S., new training and re-skilling courses and programs are being developed, including the Clean Energy Workshop Training Program in California, the Green Jobs Advisory Council in Washington, D.C., the Green Corps in Chicago, the Opportunity Austin 2.0 in Austin, TX, and the Renewable and Sustainability Degree from Illinois State University, to name a few.

Market Transformation

Governments can accelerate market transformation. Australia has implemented a new type of lease for their government occupied buildings. It is a "Green Lease." It contains mutual obligations for tenants and owners of office buildings to achieve energy efficiency targets. This improves energy efficiency by setting a minimum ongoing operational building energy performance standard.

The State of Missouri in the U.S. and many others have created Green Sales Tax Savings Program for the purchase of many energy efficient appliances.

The NGO Role

Non-government organizations (NGOs) can play an important part in the march to sustainability.

The Alliance to Save Energy's (ASE) Green Schools Program engages students in creating energy-saving activities in their schools, using hands-on and real-world projects. Through basic changes in the operations, maintenance and individual behavior, Green Schools has achieved reductions in energy use of 5 to 15 percent among participating schools. In addition, Green Schools encourages students to learn the lessons of the energy-efficiency message and apply them in their homes and communities.

The ASE Green Campus Program is leading the way towards campus sustainability by bridging the divide between students and institutional energy costs. Through Green Campus, students are working to save energy on campuses by building general campus awareness, incorporating energy conservation and efficiency into course curricula and implementing projects targeting energy use, student purchasing decisions, and operational changes. In addition, the ASE program, The Drive Smarter Challenge, addresses conservation in the transportation sector.

Organizations with the same charter as ASE are currently being formed in Australia, the European Union and India.

The United States Green Building Council (USGBC) has created a voluntary standard for building construction and modernization. The USGBC LEED Program is intended to provide third-party verification that a building or community has been designed and built with sustainability as a key guiding principal. It is now gaining international acceptance, and many regional and local regulatory policies are being modeled after the LEED Program. See Chapter 9 for a full description of the program.

Green Star is an environmental rating system for buildings, developed by the Green Building Council of Australia. It is a partner to the LEED rating scheme.

Industry's Role

Almost every major corporation in the world has announced, and is implementing, robust sustainability programs, including Dell, HP, IBM, Siemens, Wal-Mart, Phillips, United Technologies, Duke Energy, PG&E, DuPont, Dow, Bosch, EDF, Skanksa, GDF

SUEA, Acelor Mittal, Lafarge, to name just a few. A couple of examples:

Duke Energy, the Charlotte, NC, based utility has a program to help their consumers save energy, "Save a Watt." Many other large power generators/utilities around the world have similar programs. The fact that those, whose business is to create and transport power, are now encouraging energy efficiency is proof in and of itself that the march to sustainability is underway.

Another is Siemens AG, the global powerhouse in electronics and electrical engineering, operating in the industry, energy and healthcare sectors. It has three areas of sustainable development— environment, business and society—the cornerstones of all of its activities. In the area of environment, it is providing innovative products and solutions to improve both its own ecobalance and those of its consumers and suppliers. It has committed to improving its CO_2 performance by 20 percent, water performance by 20 percent, energy performance by 20 percent and waster performance by 15 percent by 2011.

End Use Consumers

The Boras Schools in Sweden were able to reduce their annual carbon dioxide footprint by 824,000 kg through an energy efficiency program that saved them 30 percent on their district heat costs, 8 percent on electrical power, and 22 percent on water consumption.

Sierra Nevada College in the United States recently built a Platinum LEED building which has been designed to achieve 60 percent energy savings and 65 percent water savings when benchmarked against the ASHRAE 90.1 Standard.

The University of Arts Berlin in Germany recently completed a modernization project that will reduce its annual carbon footprint by 1,180 tons and energy consumption by 28 percent.

Key Actions

Key actions that can continue the world's march towards sustainability include:

- Mandate efficiency and sustainability standards for new and existing buildings;

- Require the benchmarking and labeling of buildings and homes for efficiency and sustainability;

- Adopt policies to allow payment of energy efficiency improvements with utility bills or through property tax assessments;

- Provide end user tax incentives for energy efficiency improvements;

- Provide governmental support or subsidies to accelerate the development of new technologies;

- Adopt a holistic approach to energy efficiency and sustainability, think beyond the building and re-think the building;

- Utilize innovative business models, such as performance contracting, and incentivizing utilities to fund energy efficiency;

- Leverage existing technologies;

- Increase energy efficiency expertise through education: university, community colleges, trade schools, re-skilling programs;

- Every organization should appoint a chief sustainability officer and require its supply chain partners to have sustainability programs; and

- Do your part as an individual.

SUMMARY

The *"Perfect Storm for Sustainability and Energy Efficiency"* is here. *"Riding the Green Wave"* towards sustainability is under way. Great strides are being made in many parts of the world. Governments are providing important leadership. NGOs are supporting the efforts. Industry is providing solutions and services to the end use consumer and improving their own sustainability performances. And end use consumer behavior is modifying. That being said, two thoughts come to mind: the first, by Thomas Edison, who is quoted to have said, "Opportunity is missed by most people because it is dressed in overalls and looks like work," and the second, by Michelangelo, who is quoted to have said, "The greater danger for most of us lies not in setting our aim too high and falling short, but in setting our aim too low and achieving our mark." The majority of the work is still ahead of us. We cannot lower our goals. The cost of failure is too high.

Chapter 3

Packaging Sustainability

Shirley J. Hansen

We live in a special time, the start of the Anthropocene, in which humans have come to dominate almost all aspects of the biogeochemical processes that underwrite our planet. Now is the time for all students of sustainability to take the lead in understanding and applying the principles we have learned and to develop a practice of sustainability that can apply across all levels of organization from the local to the global.

Ashwani Vasishth
Associate Professor of Environmental Studies
Ramapo College of New Jersey

Sustainability is a way of thinking. It recognizes that our actions have consequences. If we are to preserve this small planet we call Earth, we must realize there can be benefits, or irreparable harm, as a result of the choices we make.

And we do have choices. If we are to make the "right" choices, we need to be informed. For that, we need leadership. An essential component of sustainability is people with effective management skills to guide us. This chapter is designed to give chief sustainability officers an awareness of the options before them, the relative advantages and disadvantages of those options, and a sense of the broader context in which we operate.

MANAGEMENT

Our energy programs have suffered for decades because we have too often failed to regard energy managers as part of *manage-*

ment. The lack of "energy" expertise at the management table has hurt us all.

It is incredibly important that we do not allow sustainability managers to suffer a similar fate. *Chief sustainability officer (CSO)* should not be a new title for the energy manager. The responsibilities of a CSO are much broader and the leadership requirements are even more demanding. Fortunately, a number of large corporations have set the stage by creating CSOs or VPs for sustainability. However, an aggressive posture is still needed if we are to effectively permeate the market place. It is truly unfortunate that most books and articles on sustainability and sustainable development do not speak to policies that create sustainability management positions.

Malcolm Unsworth, President and CEO of Itron, Inc., has declared, "… the way we *manage* the world's energy and water will shape this century." Italics supplied.[3-1]

An old adage strikes a familiar cord when we observe that if everyone is responsible, no one is. Well-meaning organization that profess to have a sustainability (or green) program are not apt to have one for long unless someone is put in a leadership position to make it happen.

Gaining Knowledge

To create a position of influence and respect, a CSO must be knowledgeable about an array of issues critical to sustainability. Much of the focus of this chapter and throughout the book is on energy, because it is at the heart of many sustainability issues and energy efficiency is the money maker for many sustainability programs.

Knowledge is more than information. You can become inundated with information from junk mail, spam, blow-ins, etc., all of which ply you with information you don't want and don't need. Knowledge demands that information be sorted through, analyzed, assimilated and applied judiciously.

This has never been more true than in these days of "greenwash." You can easily drown in a sea of "green." Those who purport to have green products or have green charlatans hiding in the bushes with all kinds of worthless advice, are bombarding us from all sides.

A key chal-
lenge to the CSO is
to develop reliable
sources in matters
of vital interest to
his/her sustainabil-
ity efforts. We don't
need to reinvent the
wheel to sort out
this mess. There are
several steps that
can be taken to save
time and effort.

**Step 1. Deter-
mine what you
need to know.**
What areas are

THE GREEN CHARLATAN

of vital interest to top management in your organization as you pursue sustainability? What topics are basic to your needs? Write it down. Prioritize it. Do not be swayed by what "they" say you should know.

Start with what you need to know to do your job effec-
tively. If you are in Las Vegas, you need to know about those water intakes in Lake Mead. In Seattle, it might be the level of snowpack in the mountains. In New York, you probably have serious concerns about future garbage disposal and could care less about snowpack in the Cascades or water levels in Lake Mead.

Our needs are different and the relative importance of those needs varies widely. We all need to be aware of the amount of fossil fuels we are importing from Canada, but this can be of far more importance to folks in Minnesota than those in Virginia. Further, knowing how much we rely on the Middle East and Venezuela might be more critical than the Canadian numbers.

So a very basic step that precedes all others is to know your job, your own organization and the conditions within which you must operate.

Step 2. List some criteria that you need to sort essential information. Your criteria for seeking advice and guidance will vary from what you need to know about specific products or pieces of equipment. Basic criteria will include the reliability of the information, how germane the data are to your issue(s), how current the data are and how useful are the depths and form in which they are available.

Step 3. Identify sources you can trust. The single best source of information is apt to be the network you develop with others who are charged with similar sustainability responsibilities. Also, it will help to identify some on-line and print media you can trust. Consider what organizations exist that can provide you with solid information, such as the Sustainability Consortium—a group dedicated to identifying *real* green products. Check to be sure an organization has integrity. Be leery of those groups that have an axe to grind; i.e., a political position that is apt to bias all the data they provide.

A fundamental concept put forth in this book is the importance of CSOs conscientiously exercising leadership and having the knowledge base to back it up. Unfortunately, there is just too much information. To get on top of the problem requires a knowledge of sources, critical thinking skills to evaluate the value of those sources as well as the ability to discern the value of the information offered.

Analyzing Sources
Information surrounds us. It is typically biased—often subtly—but biased nevertheless. Nearly every group, every individu-

al, has a bias. An obvious example can be found in the Electrification Coalition's efforts.

On behalf of the Electrification Coalition, Davis W. Crane stated at a 2009 Senate hearing: "Electrification of our transport sector is, of course, not just a major step forward in our effort to 'decarbonize' American society, it means much more in terms of the enhancement of national security and the preservation of national wealth."[3-2]

Mr. Crane's statement is pretty far reaching. Are we talking about transportation that generates its own electricity, or are we referring to transport that is plugged in? What if the "plug" is connected to a coal burning plant? How much coal and emissions are we talking about in this effort to "decarbonize" America? When we consider the 11,600 Btu value of a kWh at generation is reduced to only 3,413 at the plug (a loss of roughly 2/3rds of the power through transmission and distribution), does this really represent a major environmental step forward? Not necessarily!

SELF GENERATING

While the Electrification Coalition is striving to serve some admirable goals, the CSO needs to look beyond the face value of such declarations. If we dig a bit deeper, we find the Coalition's Mission Statement says:

> The Electrification Coalition is a nonpartisan, not-for-profit group of business leaders committed to promoting policies and actions that facilitate the deployment of electric vehicles on a mass scale in order to combat the economic, environmental, and national security dangers caused by our nation's dependence on petroleum.
>
> The Coalition seeks to achieve its goals through a combination of public policy research and the education of policymakers, opinion leaders, and the public. Equipped with exceptional research and analysis, these prominent business executives bring credibility, insight, and objectivity to the debate over electrification.

Part of assessing such statements is a look at the source. In this case, the membership of the organization. Then, the question must be asked, do the coalition's "business leaders" find it in their best interest to support the organization's position. If one wishes to gamble a bit, it is a pretty safe bet that, within the Electrification Coalition, electrical vehicle manufacturers are more apt to be represented in the group than those who drill oil wells.

In seeking out the knowledge you need, keep in mind your position of influence and respect ride on each and every morsel you pass along. It only takes one really bad piece of data to kill your credibility. If you have not thoroughly vetted a piece of information, qualify it. Include your source and any doubts or concerns you have regarding the data, the source, and/or the context within which it has been offered.

If you find that a source you have trusted has provided you with misleading information, do not hesitate to make it known to your information recipients as well as those who provided the information. It will only lessen your credibility if you attempt to ignore or cover up this information. It really hurts if the correct information comes from another source.

JUDGMENT CALLS

A basic aspect of sustainability management is leadership. Changing people's attitudes and behavior is at the heart of changing the way we think about our planet and our actions. Effective sustainability management requires some very demanding leadership skills. One of those is the ability to make judgment calls.

A peripheral issue can help illustrate the point. Your organization has many people devoted to some "green" behavior based on misinformation. "Jane in accounting" is exceedingly proud of the garden she is growing and how it contributes to a green world. She would be horrified to learn that all she has planted there has been "genetically engineered." What would it benefit to tell her? Does it really matter that her concerns regarding "genetic engineering" are based primarily on a marketing ploy to help protect farm produce in Europe?

If we set aside Jane's problem and refocus our misinformation dilemma on broader concerns within the organization's efforts to achieve sustainability, a number of issues emerge. You will be required to make judgment calls in every direction you turn. If a department is doing the right thing for the wrong reason, does it matter? Is there any point in weighing in on a heated discussion among parties in the organization on a matter, such as global warming, especially if its relevance to the sustainability goals you are trying to achieve are not tenable.

As others demand you make certain judgment calls, it will help if the specific issue has been thought through first. Consider: What are the outcomes you are hoping to achieve? Will the resolution of a particular issue at hand contribute to achieving those goals? Does misinformation in the organization on some aspect of sustainability have a significant impact on the organization's sustainability efforts? If you ignore the misinformation, could it come back to bite you? What if that misinformation is continually raised at the management table and it tends to reduce the commitment top management has toward sustainability?

The hardest judgment calls fall in the "shades of grey" category. That's why CSOs get paid all those big bucks!

Repeatedly, major judgment calls will force you to determine where to "draw the line." Typically, a concern repeatedly thrust on CSOs is the range of change needed and the timing. How sweeping should the change be and how rapidly should it be implemented? This will vary with the need, the issue and the temperament of the organization. Sometimes small, incremental steps will be most effective. Other times, a strong decisive step will be more apt to get the job done. Procedures will vary with the perceived resistance to change and how dramatic the change will seem. Critical to your efforts will be the extent to which top management is backing you, or pushing you, for that particular change.

Unfortunately, there is no magic bullet, no formula, no handbook to guide the calls that must be made. It is one of the key pieces of leadership that goes with the territory. The fact that the territory is frequently in the uncharted waters of sustainability makes it that much more challenging.

The old line about not biting off more that you can chew is particularly apt. The limits might be set by the number of people affected, their attitudes toward change, the cost of that change, or the degree of change you seek. The trick is to know how much you can comfortably chew—and swallow.

Assessing potential outcomes before you act can give you some protection. One exercise that will help take the sting out is to envision what the situation will be should things go horribly wrong. Look at it from all sides. What, if any, support base will you still have if things go sour? Imagine what the worst result could be. Can you live with it? Is it worth the risk? Test market the tragedies. If you have mentally lived through it and your head is still on your shoulders, go for it.

All of this leads to the critical importance of a master plan as discussed in Chapter 10. Master planning pushes you and your colleagues to think through many issues. Getting a "sign off" from top management becomes a security blanket worth its weight in gold.

Woven through the knowledge base and the judgment calls are several key topics that you will have to live with every day. Several of those that are most prevalent and pernicious are addressed throughout the book.

COMMUNICATIONS

Quality communications is an essential component of leadership. The ability to inform, advise and persuade are vital to getting the CSO's job done. Changing people's behavior is not easy and a large part of making it happen are effectively orchestrated communications strategies. Effective communication does not just happen... it is *planned*.

There are many fine books and courses available on the art of effective communication, the point here is to make you aware of the importance of planned communication strategies if you are to achieve your goals.

Consider internal communications as marketing. Know your targeted demographic, tailor your message and timing. It may be the same information, but the manner in which it is presented may vary tremendously from a formal presentation to the board to a relatively informal conservation in the boiler room. It does not take long working in the facilities area to realize that the folks in the boiler room talk an entirely different language from those in the board room.

BOILER ROOM TO BOARD ROOM

As you become increasingly sensitive to THEIR concerns, you will do a better job of making sure your message falls on receptive ears.

Of course, an effective way to gain respect is to have something worthwhile to say. This means a first order of business is to do the homework; be knowledgeable about key issues and be sensitive as to how much your audience needs, or wants, to know.

Position of Authority

A place at the management table brings with it a lot more than a chair and a nameplate. A seat at the management table tells the organization that sustainability is important. It offers you visible and implied support. It also offers you a venue to be heard by top management and the board. It's imperative that it not be wasted!

To effect change, the CSO must be in a position of authority. It is incredibly hard to change the behavior of those who outrank you.

You have the opportunity to enlarge (or diminish) the authority granted you by what, when, where and even why you communicate. You (and your position) are given authority; you earn respect. Others will learn quickly the value to place on what you have to say and write.

A major responsibility for any department chief is to sell the various programs under his/her supervision. In the case of the CSO, an important challenge is helping top management to understand the role energy and the energy manager play in the broad sweep of sustainability. A good place to start is to determine how the CEO, CFO and COO view the energy manager's contributions within the organization. If they do not see him/her as part of management, the organization and your sustainability efforts are missing a huge opportunity. If they don't see him/her as a source of information about energy supply—availability and pricing trends— the energy manager is not fulfilling the position's potential. It is imperative that those who hold any position in energy services become a reliable information source.

Becoming a respected part of management and a credible

source of information must be earned. This is particularly difficult for those energy managers, who became pigeon-holed in the 70s and 80s as tech geeks. This is not to say that we do not need the technicians. We need them now more than ever. But we need managers, who manage energy needs and make a strong contribution to the sustainability program.

Of course, the same can be said for others in your operation, such as those who manage water, wastewater, etc. A trickier part of managing the sustainability program might arise when you sense the need to modify behavior in another department. Consider the folks in purchasing—how tuned in are they to buying green. Or the folks in IT that need to do a better job of managing the data spewing out of smart meters. If they are not sufficiently supporting your operation, how do you work with other department heads to make it happen?

Tools of the Trade

One of the toughest parts of analyzing and planning your communication efforts is to accept that there is room for you to improve. We all like to think we are good communicators—we know how to talk, write, read, listen, etc. It's part of who we are. It's basic as to how others see us. It's hard to admit we could do it better. To consciously attempt to improve our communications skills is a struggle. It is like trying to read a book on how we read. The process keeps getting in the way of the product.

Let's start with our weakest link: listening. We learn more by listening than by reading. Unfortunately, schools seldom teach us to listen. In fact, our listening skills get progressively worse though the school years.

When I mentioned orchestrating a communication strategy, in all likelihood your mind jumped to holding meetings, developing papers, or handing out flyers. It would be very strange indeed if your first thought was about who you should listen to.

To hear is not to listen. Listening is a skill. We can be taught and we can learn to listen effectively. A CSO must listen. In turn, it important to realize how other people listen to you. There are many tricks to effective speaking—phrasing, changing the volume of your voice, deliberate pauses, etc. A little attention to how you use your voice as an instrument of communication can pay big dividends. And then there are other ways...

My grandmother, a pioneer in Washington State, used to say, "if you don't toot your own horn, no one will know you can play." CSOs take heart, Grandma was right.

A book could be written on the various aspects of sustainability training. The need is obvious, the financial benefits are well documented. The training, however, is typically focused on specific resources; e.g., energy management, water management, etc. These knowledge-based approaches are not the only way to reach the rank and file. Subtle behavior modification can yield big dividends. The magic elixir of behavior modification is recognition.

At the CSO's finger tips is an approach laced with behavior modification that would make Grandma proud. Demarcation celebrations, such as the xxx,xxx,xxx,xxx avoided CO_2 emissions, offer a great way to acknowledge some goals that have been met, recognize what individuals have accomplished, and do a bit of subtle self-congratulations.

A more structured, but valuable, way to achieve many of the same ends is to prepare reports.

REPORTING

Politicians seem to get more wrong than right these days, but one of them once told me in D.C. that is critical to occasionally, "point with pride and view with alarm." From Washington State to Washington, D.C., that's a lot of wisdom.

Reports, statements, news releases, etc. are often tedious to write, but they are essential components of the job. Done right, they can point with pride, view with alarm, and make Grandma proud.

Virtually every communiqué is designed to engage someone in your efforts. For maximum effectiveness, there are a few simple steps to follow:

1) Identify your audience;
2) Determine what you want them to know and assess their readiness to hear your message;
3) Resolve why they need to know; and
4) Decide the most effective way to get the word across.

A really good sustainability report does not write itself. It takes time to gather the resources from some rather disparate disciplines, organize it effectively and communicate the findings appropriately.

Fortunately some good help is available. The EcoStrategy Group recently compiled a report, "Trends in Sustainability Reporting: A Close-up Look at Bay Area Companies." In May 2010, Karen Janowski and Kathleen Gilligan also offered some guidance, "How to Produce a Top-Notch Sustainability Report."

The collective wisdom from these and other documents provides a few tips to consider:

a. See your report as a sales tool. Remember to "point with pride" where appropriate.
b. Give credit where credit is due. Include pictures that illustrate accomplishments.
c. Recognize where senior management's interests lie and give credit without pandering to their program support.
d. Benchmark goals and achievements. Provide enough data to support conclusions, but avoid weighing down the narrative with statistics.
e. Understand your audience(s). Use it to guide the nature and form of the presentation. Consider more than one version of the report to meet specific audience needs.
f. Use resources at hand. Do not reinvent the wheel. Whenever possible, save time and energy by using existing frameworks … adapted as needed, credited as appropriate.
g. Tie reports to other organization and community links. Practice what you preach. Keep the report succinct and digital. An inch-thick bound report does not speak to sustainability.

The worst thing you can do is make a report look like you were required to write it. That may be the case, but you might as well get all the mileage you can from it. If you see it as a challenge to bring others on board, you may not only accomplish that, but enjoy the chase. Consider that instead of a conclusion to a project, your report can be the beginning. Or, it may be a transition to a new phase or facet.

MEASURES OF SUCCESS

As it has often been stated, "You can't manage what you can't measure." As you move into new territory you will be challenged. Measurement puts certainty into uncertainty. CSO's response options to the board or releases to the public must rely on solid data, which indicate progress towards stated objectives. Either formally or informally every aspect of a sustainability program needs to be measured.

Responses require hard data. Every sustainability program needs to rely on broadly accepted measurement protocols. Energy savings protocols represent a more advanced field of measurement, perhaps because money often changes hands based on documented energy saved. To measure energy efficiency savings, the International Performance Measurement and Verification Protocol (IPMVP) is recommended. It is published and maintained by EVO and can be found at evo-world.org. [3-3]

The American Society of Heating, Refrigeration and Air Conditioning Engineers (ASHRAE) has developed a companion standard (ASHRAE 14) which provides more technical information. The Federal Energy Management Program Office (FEMP), U.S. Department of Energy, has developed a "cookbook" based on IPMVP that gives explicit directions as to the IPMVP options to be used with specific measures.

Measuring energy savings is an important step towards gathering data on emissions reduction. Once the energy reduction has been measured the avoided greenhouse gas emissions can be documented. Emission reduction information should be included in any reports. We know we can save money while cutting pollution levels, but it is great PR to underscore the multiple benefits.

Methodologies and standards for measuring and reporting water-related factors is nascent. An in-depth review of existing and emerging water accounting methodologies and tools for use in the private sector can be found in a public draft of "Corporate Water Accounting: An analysis of Methods and Tools for Measuring Water Use and Its Impacts" by Jason Morrison *et al.* [3-4]

A Dutch-based international organization, the Water Footprint Network, helps individuals and corporations get a better understanding of how and where water is used. The organization has recently released a living document, *Water Footprint Manual*, which defines "water footprint" as a "spatial and temporal indicator of direct and indirect freshwater use" and sets forth methods of water footprint accounting. It covers everything from calculating water footprints to water sustainability assessment and response options.

Other measures will be needed, often unique to your operation. A determination of whether your wastewater treatment meets certain standards may be needed. Some measure of your air changes

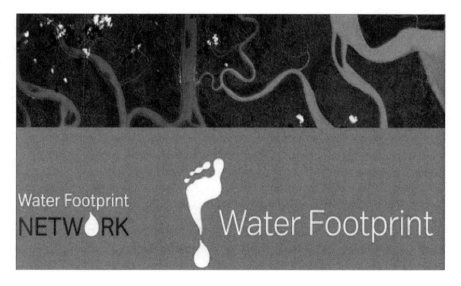

per hour and the quality of indoor air, such as CO_2 levels, could be important. Attempts to assess the extent to which certain products and services meet your sustainability standards will be ongoing.

Since you cannot manage what you cannot measure, sustainability *management* will demand a means of measuring a broad array of program activities. Some will be less formal—even casual judgments as you work through your day.

Repeatedly people measure what they "should," but fail to take advantage of what the data tell them. A good sustainability program must include routine procedures for turning data into information and information into results. David Matacek once gave a speech at the National Association of Energy Service Companies entitled, "Information is Power; Applied Information is Profit." There is a strong message in those few words for CSOs, even those who work for "non-profits."

PERNICIOUS TOPICS

There is a battery of information that CSOs must know. One topic you cannot escape is finance. Your budget, the organization's budget, the availability of outside grants, the cost of achieving certain sustainability goals will dog your heels every day. Money concerns cut across everything you do, want to do, and can't do.

Economy

For those who connect the term, "sustainability" with global warming, economics seems a secondary consideration. Concerns for the future of our planet overshadow the impact certain measures might have on the pocketbook. However, no matter how serious your concerns might be, you will ultimately hit a financial wall. As you progress from rhetoric to realities to results, there are dollar signs every step of the way. Ignoring the economics of what you wish to accomplish is an exercise in frustration. Acknowledging the economic implications of the actions you wish to take is more apt to make them a reality.

CSOs do not have the luxury of ignoring economics. While many institutions, municipalities and corporations are increasingly concerned about the impact their operations have on the planet, they all have budgets they must consider. The cost of certain measures must be weighed as key selection criteria. In fact, costs need to be judged beyond the measure and assessed in terms of the best use of the funds available—both in reference to a sustainable environment and the financial needs of the respective organization.

Government intervention almost always has an economic impact. Grants and subsidies have helped to spur renewable development. Conversely, some government actions work against sustainability. Energy subsidies keep prices down and tend to discourage energy efficiency. Government aided generation, such as the Tennessee Valley Authority and the Bonneville Power Administration, have had the same effect. Those of us who live in the Pacific Northwest and benefit from these programs, seldom appreciate that taxpayers across the country are helping keep our electric rates relatively low. Nor, do we realize that these "benefits" have worked against the need for us to use energy more wisely.

Raising energy prices through taxation is not the simple remedy many would have us believe. Higher prices prompt us to use less; however, higher energy costs impact the prices of all goods, which make us less competitive in the global market place. Ultimately, they hurt our economy and put people out of work. Ironically, the end result can be a recession, which certainly reduces energy consumption. The price to save energy by this means is painfully high.

Regulations, or the prospect of regulatory action, may discourage energy efficiency or other sustainability acts. Historically, governments have often issued sweeping regulations calling for an entire economic sector to cut energy consumption by a certain percentage. No consideration is given to the organizations that have practiced good stewardship. They are still expected to take an equal hit. A good example of this is Dow Chemical, which has been recognized as a green company, but is expected to cut energy consumption by the same percentage as other industries in its sector even though those other companies may have paid far less attention to environmental matters. The same concerns apply to all federal agencies, which are expected to reduce energy consumption by a specific percentage no matter what has been accomplished in the past.

When organizations are repeatedly subjected to such action, there is a very real temptation to relax sustainability efforts. There is a tendency to build in some "slack" in the operation to cushion the next government action. CSOs need to keep their government relations people aware of the potential impact certain pending government actions might have. In turn, it is hoped that the Congress and government agencies, once made aware of this potential impact, will be persuaded to use more judgment in the enactment of such laws and regulations.

A great deal of press has been devoted to the relatively high price of renewables, and by implication the cost of sustainability. This can be misleading. Many measures used to achieve sustainability can actually save money. The basic concept of sustainability is to do the best we can with what we've got... a rather frugal approach.

THE CREDIBILITY GAP

In addition to the trauma and drama surrounding "climate gate," there are two other arenas that will plague a CSO's efforts to engage in sustainability dialogues. First, there are repeated efforts to suggest "clean" energy can provide all the energy we need in the immediate future. As desirable as that might be, this is not even a possibility in *the long term*, much less the immediate future. The numbers just aren't there. By 2030, it is predicted that the world will need 28 million megawatts to meet our energy needs. It is also estimated that we can only safely produce about one half that, or 14 million megawatts, of carbon-based energy. That leaves another 14 million megawatts of non-carbon power we must find. We simply cannot get non-carbon based energy to that scale fast enough.

This would require massive amounts of new infrastructure, time and capital. Recent events indicate those resources are simply not attainable.

The second issue is closely related. Wishful thinkers see renewables filling the gap *vis-à-vis* energy efficiency (EE). In 2009, the venture capitalist (VC) industry invested more than $2.6 billion in supply-side technologies while the VC EE investments were roughly $440 million. Our biggest hope in the supply-side of the equations is that the smart grid will help us identify the roughly two-thirds of generated electricity that we lose each year through transmission and distribution (T&D). It is realistic to expect we can recover some T&D losses via the smart grid, but not all.

The most effective way to fill the supply gap is simply to use less. Estimates of our ability to fill the supply gap through EE go as high as 85 percent. Once we take off our rose-colored glasses, there is still an attractive aspect to EE; it's relatively cheap. It has been said before, but it is worth repeating, "The cheapest, cleanest kWh of energy is the one not generated." EE delivers "green" most cost-effectively. That is something we can all agree upon.

The problem is EE is not sexy. Renewables attract far more attention from VC investors and government incentive programs. EE has been around for decades. It is old hat. The EE financial incentives are harder to come by, but they do exist. State and federal funds are available. If we are to come close to closing the gap in future energy demands, EE is going to require much more attention.

To help us get there, the energy service company (ESCO) industry offers a lot of promise. ESCOs guarantee that the energy savings yielded from a project will pay the project costs. ESCOs not only reduce the customer's operating costs, they do it without any front-end capital expenditures from the client. What a beautiful solution! Future energy savings reduce the energy budget and make the customer and the economy more competitive while reducing emissions—while we fill the energy/credibility gap.

As we work to bridge a multitude of credibility gaps, it pays to take a hard look at one more—those who claim to be "green" and don't even make it to chartreuse.

BUYING GREEN

Buying "green" is far more complicated than going to K-Mart for a new widget.

Worries about the economy flow naturally into concerns about what we buy to support our sustainability goals. It is not as simple as it appears. First, we are mired in a myriad of "eco-friendly" companies with unfounded, often exaggerated, green claims. We also have those who press for green changes that are not really warranted. Sorting all this out, buying the right thing at the right time from companies that are truly "green" is not easy. It requires careful inquiry, enough data to question claims, and the fortitude to say, "No!" when needed.

If, for example, in demonstrating our zeal to be "green," we decide to replace a serviceable floor with bamboo. The fact that bamboo is "green" does not necessarily make this a sustainable action if, in the process, perfectly useable floors are discarded, and energy is expended to remove and dispose of the old floor and acquire and install the new. The bamboo is harvested unnecessarily, and the floors are manufactured and installed unnecessarily. The total process can consume huge amounts of energy and resources. Looking at single measures tends to overlook this perspective.

We are most effective within our sphere of influence, but we cannot ignore the rest of the world. It will intrude. Our actions affect the world around us, but others can impact it as well. Being "green" interjects the world into the process.

Take something as simple as my socks, which are 77 percent bamboo. They have a wonderful softness and I like wearing them. Are they, however, better for the environment than silk? How much natural resources are consumed growing bamboo compared to the efforts of the industrious little silk worms? How much more energy is required to make the rayon from bamboo? If the world suddenly favored bamboo over silk, what would be the economic impact on Japan?

In addition to the actual product, "green" packaging is moving front and center. A major effort is being made by many firms to

minimize their packaging or they are switching to different, more eco-friendly materials. Kraft has shed packaging across a whole range of items and Mountaire Farms has removed the wax coating from its boxes so they can be recycled. In a "shades of yesteryear" move, the Straus Family Creamery has gone to using glass bottles, which are reused by the creamery.

Some ideas are subtle and fun while others have the force of the pocketbook behind them. On the fun side, *Opportunity Houston* magazine sent out a postcard inviting the recipients to view its spring 2010 "Green Issue" in digital format. But that's not all, once the recipient has used the digital address from the card, the reader was then encouraged to plant the card to grow an array of wildflowers, "… thus completing the green cycle." At the other end of the spectrum, Kaiser Permanente now requires its medical equipment and product suppliers to provide environmental data on all supplies; thus, affecting more than $1 billion in goods each year.

As new products come on the market, it is hard to sort out

those that you can rely on to deliver on the claims. In addition to checking out the product, it helps to check out the company making the claims. If the company has documented sustainability practices, the odds go up that it is offering a quality product.

The Powerhouse Solar Shingle is a good example. It was named one of the 50 "best inventions" by *TIME* magazine in 2009. This third party recognition suggests it might be worth further research. The Powerhouse Solar Shingle can be integrated into rooftops with standard asphalt shingle materials and can be installed simultaneously by roofing contractors without any additional training.

A hard look at the Powerhouse Solar Shingle's manufacturer, Dow Chemical, can give us an indication of the credibility of its claims. According to its 2009 annual report, Dow has reduced its "Kyoto" emissions by 35 percent—exceeding the Kyoto protocol, it has underway plans to build an algae-based bio-refinery in Freeport, Texas, which will convert CO_2 into ethanol. Further, the company, which is the greatest consumer of energy in the U.S., is on-track to fulfill its commitment to reduce its energy intensity by 25 percent. This kind of sustainability track record can give CSOs more confidence in a product. It also makes one want to support a company that demonstrates this level of social responsibility.

One resource to help CSOs sort this all out is the Sustainability Consortium, which is headed by Jay Golden and Jon Johnson, who are respectively from the Arizona State University and the University of Arkansas. Johnson describes the Consortium's task as, "... creating a system that would enable companies to get information on product categories or products."

The Sustainability Consortium is an independent organization of diverse global participants, who work collaboratively to build a scientific foundation to drive innovation to improve consumer product sustainability through all stages of a product's life cycle. Thus, the Consortium provides some screening and also promotes new products to meet sustainability needs.

Rather inconsequential decisions we make can have repercussions far from our shores and impact the sustainability of our planet. Further, the socks dilemma underscores a familiar theme. We simply don't know enough about the environmental consequences of many of our decisions. Of course, one can "Google up" all the unknowns and, hopefully, gather enough accurate information to make an *informed decision*. But look around. Every day we are faced with hundreds of small decisions and seldom care enough to ask the right questions—or get the answers.

There is a serious question, however, as to whether we should. If we did the research each time we are confronted with such choices, we'd very likely become paralyzed into a non-decision mode. Where is the line on sufficient information?

Another supplier we all use, utilities, may come in varying shades of green. Their primary source of fuel (which may vary with the time of day) may become a key factor in your energy shopping. If your utility provides you power generated from oil from 2:00a to 6:00a and uses wind from 8:00p to midnight and you have the flexibility to shape your demand profile, you may want to opt for the wind power.

As renewables become a more important part of intermittent power, we will need to find ways to manage the changes in utility demands. As things progress, it is expected that the more sophisticated demand response, load shaping along with energy storage, will be needed to integrate all the emerging renewables without bringing down the grid. As a CSO, it will help your operation as well as the utility's if you make it a point to find out what your utility is currently doing and its plans for the future.

Coming rapidly to a utility near you is the whole new world of Smart Grids and Smart Meters. It is expected that the load management opportunities offered through these technologies will even out capacity issues for utilities, which in turn could reduce the need for additional generation. According to the Federal Energy Regulatory Commission (FERC), time-based rates have the potential to reduce peak rates 5 to 12 percent. It is also expected to provide real time consumption information, which typically inspires customers to use less energy. With these benefits, the CSO is apt to be confronted with a whole realm of new billing procedures, rates and availability concerns.

GIRDING FOR THE GRID

The advent of the smart grid is a perfect example of *management meets technology* for CSOs. While understanding some of the technologies and opportunities afforded by the new grid and smart meters are key, the true challenge will be in the finessing the organization's position to take advantage of this new world.

Utilities have changed terminology from the old "rate payers" to "customer." Now, smart grid operations are expected to force a restructuring of the industry, the individual utility and how customers are regarded and served. Utilities are moving toward the concept of "valued partner."

Allan and Chistensen, writing for the May/June, 2010, issue of *EnergyBiz*, draw on their Accenture experiences in 75 smart grid projects worldwide to observe that, as utilities move to a smarter grid, the move is expected to impact "… some 70 to 80 percent of utility business processes. As a result, focusing on just the technology aspects of these major initiatives may delay or negatively impact benefits realization, not only for the utility but also for the customer and the environment." [3-5]

They go on to note the need to align leadership vision by "… designing business processes that support a new operational culture." They foresee that over the coming decades these smart technologies will drive three key operational shifts to:

- Improved operational efficiency and resilience;

- Integration of renewable and distributed generation; and

- Unprecedented empowerment of customers.[3-6]

Of great significance to CSOs is the "unprecedented empowerment of customers." The smart grid is expected to generate a volume of data several orders of magnitude over current levels. The incredible amount of data and information that will flow from utility to customer and back can indeed empower customers. It can also empower utilities and hackers. Utilities are not apt to give up control easily. The CSO's challenge will be to have timely access to actionable data. CSOs will also need to monitor this evolving relationship and use it to the organization's advantage.

Utilities are being advised to put customers at the heart of all they do by incorporating customers' desires into their operation and business decisions. This could amount to "red carpet" treatment for CSOs to convey those desires to the utility. After all, they cannot cater to your organization's needs unless it has been made known what you need—and want.

Similar to those who used to sense how things were going in the war effort by the number of pizzas delivered to the White House, there is a very real possibility that outside data gatherers will be able to tell others more about your operation than you want them to know. Security and the protection of data could become major concerns. This introduces a complexity that underscores the need for the CSO to work closely with the chief information officer and others as he/she works to support sustainability in its many manifestations.

Closely aligned with smart grid and energy efficiency opportunities are the opportunities utilities offer customers through demand response (DR). Proponents of demand response programs cite numerous benefits of such options, including improved system reliability, cost avoidance, greater market efficiency, reduced negative environmental impacts, and better customer service. Driven by a growing demand for electricity and increasing energy costs, Pike Research forecasts that the DR market will expand to comprise 62,500 megawatts under management by 2015. For CSOs the opportunity rests in combining energy management and automation systems with an IT interface.

MAKING THE BUSINESS CASE

We've all heard it: "Get top level buy in!" If the folks at the top think something is important, you are more apt to get the attention of the rank and file. Making a business case for sustainability, or its core effort—energy efficiency—can go a long way toward getting the executive suite to buy in.

Making the most of energy efficiency opportunities is definitely a key part of sustainability management. A very basic tenet is: ***Energy efficiency is an investment; not an expense***. Energy efficiency (EE) is a *very sound* investment. We talk rather glibly about 2-year paybacks, but it's hard to find 50 percent return on investment. EE is an incredibly cost-effective way to cut operating costs; and, through such financing options as performance contracting, cut those costs without any up-front capital expenditures. As an added plus, EE can deliver the money to carry much of the sustainability program.

Reducing operating costs is good for the customer, the market, and the country's economy. Further, studies have shown that EE creates five times as many jobs per megawatt hour as does the creation of new generation.

Effective energy management should be a vital part of a sustainability program. To make it happen, CSOs have two critical, but very different, "audiences": the facility/O&M group and top management. A key component of effectively marketing sustainability is being sensitive to the audience, its interests, needs, attitude, etc.

CSOs are bound to hear a lot about our environmental needs, but alarmingly little about the most cost-effective way to meet such

TOUGH SELL

needs. People should be lined up at the door; eager to invest in energy efficiency. But ironically, it's a tough sell. It seems we need a better way to convince top management and staff of the tremendous benefits EE delivers.

Many years ago, a dear friend, who was in top management of a major corporation, gave me some sage advice, fundamental to our problem. He said, "Shirley, you folks must learn to fish from the fish's point of view." Good idea, but first we must consider the fish we are trying to catch.

The Anatomy of an Audience

For many of us over the years, it was a constant exercise in frustration to try to sell energy efficiency to the CEO. We thought we were turned away due to the discomfort top management felt when the subject of "energy" was introduced. It certainly played a part, but in retrospect we now realize that other concerns, often more important concerns, were at play.

The horrible truth is that top management is not particularly interested in *ENERGY!* They don't want to hear about British thermal units, or kilowatt hours. Finally, we have figured it out: CEOs and CFOs **do not buy energy;** *they buy what it can do.* We simply cannot get them excited about using something more efficiently, which is basically non-existent in their lives. The only time energy seems to reach their consciousness is when there is a shortage or a sudden power outage. And even that attention seems to be short-lived.

Unless we are careful, boiler room folks will see an EE vendor as part of the problem; not the solution.

Against a backdrop of management that doesn't care and facility people who have every reason to resent the "sales" job, it is no wonder that energy efficiency is a tough sell.

Fishing in the Facility Pond

"Catching" the facility people is critical for two reasons. They may not be the ones who give the top nod to a sustainability effort, but they can make the CSO's life miserable. We should never lose sight of the fact that operations and maintenance (O&M) people are the ones that turn the knobs and flip the switches.

If the effort is made to give them something *they* think they need, it's amazing what can happen. A program that meets some key O&M needs can make the difference. But it need not be tools or specific O&M measures, it may seem a little like "warm and fuzzy" stuff, but most facility people hunger for a little recognition.

O&M people are also far more important to a sustainability effort than most might realize. During the years the US DOE was providing energy grants to schools and hospitals, the Department commissioned a study to assess the effectiveness of the program. The study found that up to 80 percent of the savings could be at-

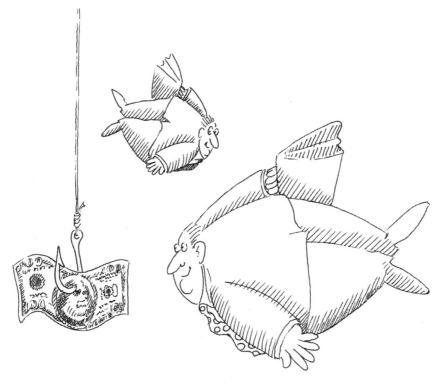

FISHING FROM THE FISH'S POINT OF VIEW

tributed to the energy efficient practices of O&M personnel ... $4 out of $5 in savings came from what the O&M folks did, not the hardware. The involvement of O&M personnel in the sustainability program can make a vital difference. It is well worth the time it takes to sell the idea to the staff.

Selling Management

To get top management's attention, money is key.

Since the money for wasted energy goes up in smoke, one way to get their attention is to literally burn money. Pile some dollars (on a fireproof tray, of course) on the desk and put a match to it. Then, remind them that money paid for wasted energy just goes up the smokestack. *Never to be recovered.*

Being regarded as an organization dedicated to sustainability can be a huge advantage. Even better, the organization can make money through energy efficiency, winning kudos as it improves the environment.

A major challenge facing CSOs is reminding top management as forcefully as possible that energy efficiency (EE) can be a self-funding endeavor. CEOs and CFOs have a tendency to compare energy investments to other business investments and fail to appreciate that no new money may be required to do the EE work. The money needed for energy investments is already in the budget—and being spent on wasted energy. The financing source for the EE investment is right there in avoided utility or power company costs. That's money that will go up the smokestack—creating more pollution each day that the energy efficiency measures are not taken.

It is always helpful to put ourselves in management's shoes as we need to point out the benefit to the bottom line. Industrial managers typically view energy as a raw material needed to produce a unit of product or deliver a particular service. If 30 percent of a cement factory's raw material costs go for energy supplies—and if you can reduce that amount by 20 percent, you can show management the direct EE benefit of 6 percent in the bottom line. That's huge!

Then, as all good fishermen know, you have to "reel them in." So ask them to tell you of another approach that will bring 6 percent to the bottom line. Work through the options with them. You've got a winner—and pretty soon they will realize it!

In industrial settings, energy efficiency is a particularly hard sell because an investment in production has a better fit with management's way of doing business. More production and shiny new equipment make their hearts sing.

It is really quite simple to set up an energy efficiency vs. increased production worksheet comparing monetary savings and associated investments on an energy vs. production basis. When we add in the costs of packaging, promoting and delivery a product with the uncertainty of a purchase, the energy efficiency component comes out looking pretty good.

Our biggest challenge may be convincing top management that EE can be a self-funding endeavor. We must be sensitive to how they perceive things. We need to remind ourselves again *and again* that CEOs and CFOs have a tendency to compare energy investments to other business investments and fail to appreciate that no new money may be required to do energy efficiency work. Money needed for energy investments is already in the budget—and being spent on wasted energy every day.

The CSO, therefore, needs to keep in mind he/she is not selling energy efficiency *per se*, but *reduced operating costs*. Reduced operating costs lead to; a) more competitive pricing in the market place; b) a bigger net profit for the company, and/or c) more money to dedicate to the organization's mission. Or d) all of the above.

The financing source for the EE investment is right there in avoided utility costs. In contrast, the money to increase production is new money that must be added to the budget, which can weaken the competitive position.

Every moment that we let economically viable energy efficiency measures go unattended we are paying out good money to pollute the air around us and damage our Earth. That makes no sense at all!

WATER MANAGEMENT

As a harbinger of things to come, a battle is now being waged on the north slope of Mt. Hood in the small town of Cascade Locks. It involves many of the players one would expect to find in a sustainability war: water scarcity, an endangered species, and environmentalists vs. a corporation. Nestles has found a lovely new spring, which yields about 100 million gallons of water annually for its Arrowhead bottled water. Aside from the environmentalists' concerns about transport energy and the billions of plastic bottles, they are concerned about a hatchery fed by the spring for Idaho Sockeye Salmon, an endangered species.

To resolve the problem, Nestles is replacing the spring water with municipal well water for a test batch of 700 fingerlings. Worried that activists might sabotage the test, the tank is under lock and key, security cameras and the watchful eye of the Oregon Department of Fish and Wildlife. In the spring of 2010, three fish in the tank died and Nestles was immediately under verbal attack. The company insists it is managing "water in a sustainable way."

When it comes to water, Cascade Locks is another pebble announcing an avalanche. We are on the brink of a global water crisis. Lloyd's 360 Risk Insight report, "Global Water Scarcity," offers a stark reminder that we are facing diminishing water supplies. The report states that only 3 percent of water in the world is fresh water and only 1 percent of that is readily usable for humans. Further, the report notes that increased populations and higher per capita water requirements (resulting from higher living standards) are all threatening this already limited supply. [3-7]

In a climate reminiscent of the 1973 Oil Embargo, we seem to be impervious to, or in denial of, the imminent crisis. Only this time the wake-up call will be water; not energy.

Water is big business. It is essential to most industrial processes. It is vital to life. The supply is not inexhaustible. Several US cities are already in trouble.

Water supply comes to us from surface water (lakes, rivers, canals), ground water (aquifers, wells), and the sea through desalination.

Water supply is the process of providing water (by self or another party) of different qualities to a range of users. In 2007, more than five billion (54%) of the people in the world had access to piped water. Another 1.3 billion (20%) relied on improved water. Roughly one billion (16%) did *not* have access to improved water and had to rely on unprotected wells, canals, rivers, lakes or springs. It is worth noting that "improved" water does not necessarily mean the water is potable.

Through great engineering creativity (or idiocy) we have Las Vegas, built in the middle of a desert using a disproportionate amount of water. The city is dependent on Lake Mead where the level of the lake is dropping. Steps are under way to access water at a lower level. These measures, however, just defer the day of judgment. Hopefully, it will buy enough time for the city leaders to figure out an alternative. Of course, if the "damn the dams" people get their way, Hoover Dam will go and take Lake Mead with it.

Word has it that El Paso is hanging on by a thread and its water supply is a constant worry.

San Diego has a water crisis, which is being met by investing heavily in the construction of a desalination plant. Desalination involves the reverse-osmosis process. Water is forced at high pressure through a filter that removes impurities. San Diego has the great advantage of living on the edge of the world's biggest reservoir, the Pacific Ocean. Ken Weinberg, the county water authority's director, summed it up, "The fact that there's no large groundwater basin limits our opportunities. We have very limited sources—you have recycling, you have conservation and you've got the ocean." It is projected that 89,600 acre-feet of the region's water supply will come from the ocean by 2020.

There is, of course, an ironic twist to San Diego's solution. It will take significant amounts of energy. So we now have water vs. energy—and the associated emissions.

Light snow packs in the Pacific Northwest and drought in the Southeast are making more people conscious of their water supply. Unfortunately, most people think the problem is temporary, and will be cured by a good snowy winter or a nice rainy season. Water problems around the world, however, are not apt to go away so easily.

While written for investor consideration, the *Murky Waters?* report[3-8] offers some valuable insights on water management by the economic sectors of beverages, chemicals, electric power, food, home building, mining, oil & gas, and semi-conductors. Water risk analysis by physical, reputational, regulatory and litigation risks provides some excellent planning information. The report also addresses water investment opportunities as it calls attention to products, such as WaterSense appliances, xeriscaping, moisture sensor irrigation systems, and the new BASF plastic filter membranes for cleaning and treating water.

Water is not only desirable for our life style of grass, parks and water fountains, it is essential to our bodies. Priorities will be required. Plans are apt to call for limiting, even restricting, water usage. It will not be easy. From the land of plenty, many will find themselves in pseudo-desert conditions. Landscaping changes and other processes that use less water will need to be considered. Technological alternatives that can reduce water consumption need to be researched and considered.

In the long haul, CSOs should anticipate progressively higher fees and growing limitations on water usage. They should also expect government mandates and regulations.

In short, our new world of sustainability will have a much different water profile than the one we enjoy today. Areas with abundant water will become the "Saudis" of the coming decades. Those with limited supply will see a drastic change in their economy, culture and way of life. The transition will place huge demands on CSOs.

Wastewater Treatment

If our emerging water supply crisis is not enough of a challenge for CSOs, they must also consider wastewater treatment. Wastewater treatment (WWT) serves two major purposes: 1) care in the disposal of waste; and 2) ways to recover water to use for non-potable purposes. WWT is both an art and a science.

In an effort to kill two birds with one stone, the City of Atlanta is in the process of powering a 122-million-gallons-per-day wastewater treatment plant (WWTP) by using cogeneration, which will use currently flared biogas as its fuel source. The WWTP is the city's second largest electricity consumer (second only to the airport). The project is expected to reduce annual operating costs by $1 million, reduce GHG by 13.7 thousand metric tons and replace 16.6 million of the 67 million kWh now required. Capping the existing biogas flares will result in reduced NO_x emissions of about 1.1 ton per year.[3-9]

Most of us do not know the difference between blue water, graywater, and blackwater. In fact, it may come as a surprise to know that they are legally defined. For example, the International Convention for the Prevention of Pollution from Ships, commonly referred to as MARPOL, defines graywater as drainage from showers, washbasins, laundry and dishwashers. The 1972 US Clean Water Act adds galley/kitchens, bath water, AC condensate, pool and spa water to the list. Blackwater is water from toilets, urinals and medical facilities. In situations where the water is treated, blackwater is collected separately from gray water as it contains potentially more harmful bacteria. Blackwater requires processing by advanced wastewater purification (AWP) systems. Oversight for such processes at sea are certified and approved by the U.S. Coast Guard or the International Maritime Organization.

AWP systems are primarily screening devices to remove large solids down to progressively smaller ones and then the water is treated by oxidizing with ozone and with ultraviolet light systems. The confines of a ship, particularly a cruise ship which is a floating hotel, offers a good illustration of what can be done to supply water and to treat wastewater. The case study below of the Celebrity Cruise ship, *Constellation*, provides an example of what can be accomplished.

CSOs need to put water supply and wastewater treatment on the check list. Determining what their respective water supply(ies) is could be the number one priority. For many CSOs, it is time to

CASE STUDY: CONSTELLATION

From the Rime of the Ancient Mariner comes the line, "Water, water everywhere and not a drop to drink." That will never be said of the Celebrity Line's *Constellation*. The ship has its own desalination plant. While the ship carries and uses a lot of fresh water, it has the desalination process as a backup. And as the 2nd Engineer, Mr. Ilias Tzanos, said, "We have very large reservoirs."

Even more remarkable is the ship's meticulous attention to its wastewater. There was a time when the trail left behind by ships was essentially an open sewer in the ocean. Today, the water *Constellation* discharges into the ocean exceeds municipal treatment standards. The ship has bilge water, ballast water, graywater and blackwater treatments.

Maritime regulations require untreated blackwater be discharge outside the 12 nautical mile line at a speed of 4 knots. *Constellation* discharges only treated blackwater outside 12 nautical miles at a speed of 6 knots.

Graywater is treated through state-of-the-art AWP systems and is tested to be sure the discharge meets, or exceeds, US Public Health Standards, according to Mr. George Kakaroubas, *Constellation's* Environmental Officer.

The *Constellation* also treats and tests its bilge-water using two oily-water separators and two oil-content meters to monitor the discharge. A "White box" is used to record information and operation details and is under lock and key—accessible only to the Chief Engineer.

Interestingly, the cruise line is also exploring its ballast water. Ships use sea water for balancing purposes. As the ship takes on and discharges sea water, there is a concern that sea life indigenous to one area might be transported and discharged somewhere else. The cruise line is trying experimental technology to advance the science of ballast water treatment and also exploring the use of water from its AWP systems.

It is no wonder that the Celebrity Cruise line, and the *Constellation* in particular, have been recipients of environmental awards. In 2009, the "Green Attitude Certificate" was awarded to the *Constellation* by the Port of Helsinki for its "high level of initiative in wastewater-related issues." The award also noted, "Celebrity *Constellation* takes its environmental responsibility seriously and sets a great example for everyone." 3-10 start thinking of water as a precious commodity. Water management needs to be included in sustainability master plans, as discussed in Chapter 10.

TRANSPORTATION

The focus for most of this book is on facilities and processes, but CSOs cannot afford to ignore the sustainability opportunities in the transportation sector. The American Automobile Association (AAA) offers some valuable tips on reducing fuel consumption and cutting emissions.

AAA points to an Australian study that stresses that it is more important how one drives a vehicle than what vehicle one drives. The Royal Automobile Company of Victoria found that a large vehicle driven conservatively can achieve better fuel economy than a smaller car driven aggressively.

In addition to pointing out the losses we incur from accelerating too rapidly from a stop, AAA also cites a Canadian Office of Energy Efficiency study on idling. The study revealed that waiting for a car to "warm up" benefits the driver more than the car. The Canadian study found that for maximum benefit, cars need no more than 30 seconds of start-up idle in the deepest winter.

Perhaps the frosting on the cake is that several studies have shown that safe drivers also get the best fuel economy. Once again, sustainability benefits are related to changes in behavior. The AAA offers some change of behavior advice useful to CSOs.

> But how do you get people to change their behavior? One way is to simply play it back to them. As experiments with household electrical use have shown, feedback is a wonderful behavioral change tool, and when companies or even cities have undergone trials in which large numbers of cars were equipped with real-time feedback devices (by companies like Cartasite or Green Road), fuel efficiency and safety have both improved. Drivers report being surprised by how much time they spent idling and how many "hard braking" events they were recording.[3-11]

THE SMALL PICTURE

While the business of managing sustainability seems over-whelming at times, there are ways to start small. A recent "School of the Future" competition sponsored by the Council of Educational Facility Planners (CEFP) makes the case. Schools submitted entries of their vision of what schools of the future would look like. Designs included solar, water, and wind power; roof gardens for cafeteria and neighborhood consumption; even underwater class-rooms. The students' sustainable vision included desks and chairs made from recycled materials. [3-12]

Providing broad horizons for school children, the CEFP competition offers a great example of how small endeavors can reap great rewards with the potential to benefit our society long into the future.

THE BIG PICTURE

Sustainability is not just an environmental issue, it is an economic transformational issue. It is a concern that reaches beyond your organization, beyond your community, even beyond your country. It is a staggeringly large issue with global implications.

Sustainability offers an open field where "green" meets economics. It brings hard realism to some of our environmental wishful thinking.

A good example of where "green" and economics cross paths, or swords, is the world of emissions credit trading. The rationale behind emissions trading is to make sure that emissions reductions take place where the cost of the reduction is the lowest. In turn, the theory goes, the overall costs of combating climate change will be kept at a minimum. This approach, unfortunately, is frequently muddied by government intervention and by opportunists ... including those in government ... that see this as a mechanism to levy more taxes.

CSOs, who see trading in their future, are encouraged to read

GHG Emissions Credit Trading. 3-13

As a companion piece to economics, a basic tenet of sustainability is to incorporate the "people factor" into a sustainability program. The management focus establishes the critical nature of leadership. The value of education and training becomes integral to the approach.

Each CSO has a span of control, usually limited to the organization or the community. The action in one sustainability program creates ripples like a stone dropped in a pond. Other sustainability programs also cause ripples, which can intersect yours in a positive reinforcing way or work against your plans.

While your actions may seem limited, your operation is part of a much bigger picture. A picture that can impact your efforts as well.

At the very least, your top management has a right to expect you to be informed as to broader sustainability needs and to be conversant with potential remedies and actions being taken by others. Recognizing the tough challenges beyond your doorstep plus their multi-dimensional nature is part of the job. A seemingly bottomless pit awaits us when we consider the growing population, the 1.6 billion people that still do not have access to electricity and the 1.4 billion who do not have clean safe water to drink. There is a prevailing and growing demand for more services. At the same time, we are experiencing a depletion and degradation of our natural resources.

These huge problems are apt to get even bigger. By 2050, the population is expected to reach 9 billion. The drift towards greater urbanization will drive a concomitant push for even more services. The pool in which we make a small ripple is apt to become really rough waters.

If we consider just the environmental aspects of a sustainability program, we need to realistically appraise existing conditions and face the harsh reality of what will happen if no action is taken. This jolt of realism will, hopefully, drive us to contemplate what needs to be done.

Chapter 4

Energy Supply Options

Shirley J. Hansen

Energy is the life blood of our economy. One cannot manage sustainability without addressing energy supply needs. It is also impossible to hold a responsible position at the management table without being informed as to an organization's energy supply options and the trade-offs they offer.

Energy supplies are a critical cost of operations—up to 30 percent of the cost of production for something like cement.

The relative emissions, and environmental implications of a certain energy supply is essential information. However, cost, availability, reliability, associated benefits and problems are all factors that also must be weighed. This chapter brings chief sustainability officers (CSOs) a series of briefings offering pertinent information on each major fuel source as well as highlights regarding certain supply options. The briefings are not intended to be exhaustive or comprehensive, but rather to offer the CSO enough information to create a global perspective and to help determine which fuels might best be used locally. And which fuels might warrant further research.

Too often people use sustainability and renewables as synonyms. CSOs must consider a broad array of concerns—much broader than energy and certainly far wider than renewable energy matters. Nevertheless, a major focus of any discussion of sustainability centers on energy usage and certainly on the sources of energy used.

Fuels are generally categorized as renewable and non-renewable, but there are shades of gray. Not everything is equally renewable. Or, for that matter, are they equally available or reliable.

Further, the costs and potential emissions from the manufacturing process are key considerations. Windmills and solar panels do not suddenly appear. They are manufactured, transported and installed. Each and every source of power has a cost and an environmental price. They are not uniform.

Most importantly, a renewable source is not necessarily a low emitting source. 30 years ago, there was a tremendous push for wood burning stoves for home heating. It was the *NATURAL* thing to do. The trees would grow back, so wood burning was, and is, considered a renewable source. Ultimately, we began to realize that wood waste would not meet demand, replacement trees grow slowly, and the environmental benefit from trees could be lost. Much to the chagrin of some nature lovers, wood burning was finally recognized as a high pollution emitting heat source. Today, for example, there are "burn bans" in the Pacific Northwest, a time, by law, when burning wood is not allowed because of the pollution potential.

Too often we assume renewables are environmentally benign. This is not the case. Conversely, there is an assumption that non-renewables will seriously damage the environment. Not necessarily so. Nuclear energy is a nonrenewable source of power, but is one of the lowest emitters among all our options. At the same time, the concerns related to nuclear waste remain unresolved.

TRANSPORT AND STORAGE

Of the many challenges facing the energy supply industry, the transport and storage of fuel supplies is a critical, and overriding, consideration. It cuts across all fuel categories and sources. One of the more troubling aspects of our energy situation is the resistance to bringing power from generating points to users. This resistance to wires and pipes stands in the way of maximizing the use of renewable.

The current focus is on eight technologies: superconducting, magnetic energy storage, thermal storage, ultra-capacitors, batter-

THERE MAY BE SOME RESISTANCE

ies, flywheels, and compressed air storage. All of these technologies are expected to compete for the electric utility energy storage market, which is expected to exceed $2.5 billion within five years.

Improved storage is expected to enhance reliability and power quality and should reduce CO_2 emissions and the need for added generation. It is also anticipated that information gleaned from the Smart Grid will help us identify and remedy losses experienced between generation and the user. Storage enhancements are greatly needed if we are to realize the full potential of utility scale solar and wind farm installations.

Smart Grid

We all know that our existing electric grid is in bad shape. The question often turns on repair or replace? Repair is not adequate and replace is hugely expensive. A 21st century grid, however, should not be more of the same. From the point of generation to where you flip the switch, it's time to apply the technology now at our finger tips. As we replace those "wires" across the nation, we can surely make any new grid more functional and more responsive to our needs.

Already we know that the grid of tomorrow will have digital technology allowing us, among other things, to have two-way communication. Imagine a grid that will let you know instantaneously how much power you are using for a particular activity. Smart Grid and Smart Meters are a management issue, as discussed in chapter 3, but this new technology also represents a supply/availability issue.

With a Smart Grid, utilities will be able to reach into neighborhoods, and, remotely and temporarily, turn off certain functions in participating facilities, such as hot water heaters. Increasing demand on the line could trigger an immediate management shut off function, which would relieve the strain on the power grid.

Many of these possibilities are being assessed at test sites with experiments in modernizing the electrical grid. One vision of how to use all the data generated by a Smart Grid is to create a business model that allows prices to fluctuate in five- to ten-minute intervals—in essence, a short-cycle peak demand. Customers will pay more when the grid becomes congested (analogous to rush hour traffic). It is expected that this effort will help avoid brownouts and blackouts as well as encourage conservation. Obviously, it will take real time pricing to a new level where costs and perhaps power sources, such as wind, can be immediately determined by the customer. Ultimately, all this may defer or avoid new generating plants and their associated pollution.

Getting a handle on the Smart Grid with all its interactive components is going to take all our smarts. Meir Shargal and Doug Houseman, in an article adapted for *Smart Grid News*, suggested a smart grid conceptual architecture, in which they identified five major components and seventeen subcomponents that will be needed to manage the effort. These numbers represent just one example of the collection of complex decisions before us. The tasks seem almost overwhelming.

Despite all the benefits being touted by Smart Grid proponents, there is growing opposition. The potential ability of utilities to reach into an organization to adjust thermostats or turn off equipment has prompted a "Big Brother" outcry. There is also concern that rates will be raised in a punitive fashion toward customers who do not have the time, inclination or sophistication to play the Smart Grid game.

Skeptics point to a myriad of problems, which could plague the successful use of the Smart Grid. The very data that are points of pride for proponents are also considered major stumbling blocks. How we manage such a huge influx of data and make them fulfill their promised potential while offering privacy protection is a subject of much debate. A model offering such incredible information technology also poses huge security issues. Intertwined with the security question is the issue of who will be driving the Smart Grid and who will own the data. Finally, there is the little matter of who is going to foot the bill. The federal government? State? Utilities? *Which, of course, all translates to tax payers/customers.* So the bottom line is *we* will pay for it, but will we profit from our investment—or merely pave the way for others to exercise more control over our daily lives and factories while they make a bundle?

The Smart Grid issues help to illustrate that transport and storage of energy is as complex and confusing as selecting the supply source that best meets an organization's needs. In weighing any fuel source, the CSO needs to assess the transport and storage situation in his/her locality. Such considerations affect reliability and price as well as impacting the environment.

RENEWABLES

Renewable energy plays an important role in the energy supply. When renewable energy sources are used, the demand for fossil fuels is reduced. In 2008, 7 percent of our total energy consumption was provided by renewables. Of that 7 percent, over half was from biomass, hydropower contributed 34 percent, wood 7 percent, geothermal 5 percent, and solar 1 percent.

Over half of renewable energy goes to producing electricity. About 9 percent of U.S. electricity was generated from renewable sources in 2008. The next largest use of renewable energy is the production of heat and steam for industrial purposes. Renewable fuels, such as ethanol, are also used for transportation and to provide heat for homes and businesses.

Wind

When we think of wind, the picturesque windmills of Greece and Holland come to mind. Windmills have been a source of power in those countries and the farm lands of America for a long, long time.

Today, we are more apt to think of wind farms in Texas or California. While wind can be an attractive source of energy in some localities, it represents a very small percentage of energy sources in the U.S. As a result of the recession, it has recently dropped in production. By 2025 wind generation is expected to reach 200,000 megawatts in the U.S. skeptics point to the unreliability of wind and the need for back-up when the wind does not blow. They also express considerable concern about the potential increase in prices. Ofgem, the power industry watch dog, says prices for wind generated energy could rise by 60 percent by 2012.

The biggest obstacle to massive use of wind power is the electronic transfer. Environmental concerns regarding wires have prevented several large wind farms from realizing their potential.

In the UK, wind power has been championed by former Prime Minister Gordon Brown and his team. It is hoped that wind will help bring balance to the UK energy mix though it only contributes 2 percent at this time.

Solar

Solar power offers users the means to capture useable energy from the sun. Solar is not new. We have relied on it in many forms for centuries.

The advent of photovoltaics has made solar a more viable source of energy. While some emissions come from manufacturing solar power equipment, solar offers a very attractive source of energy for those concerned about sustainability. With the growing demand for clean energy, the use of solar has grown dramatically. Today, solar contributes about 6.43 gigawatts to our world energy needs, up 6 percent from 2008.

When we think of solar, we envision an array of solar panels. However, advances in technology have made miniature solar cells, which power small electronic devices and microelectronics, increasingly popular for use with small scale appliances.

Solar is spreading its wings literally and figuratively. After seven years of research, testing and perseverance, the first all-solar-powered airplane flew for 87 minutes in Switzerland in April of 2010. The *Solar Impulse* had an inaugural takeoff and landing in December 2009, but this was its first true flight. The plane has a 208-foot wingspan with 12,000 silicon mono-crystalline solar cells on its wings and horizontal stabilizer. The solar cells provide power to the craft's four electric engines.

A major obstacle to the wide use of solar has been the process of getting the generated energy to the user. Transmission of large amounts of energy has been facilitated by ABB's high-voltage, direct current technology (HVDC). This technology pioneered in the 1950s by ABB reduces transfer power losses to 3 percent or less. HVDC has fostered greater use of renewables as it is also applicable to wind and hydro power.

Fuel Cells

There is a growing interest in what fuel cells might contribute to the energy mix. The large (200-400 kW) cells, such as those made commercially available by United Technologies, have been a matter of interest for some time.

The fuel cell is not a new concept. The principle of the fuel cell was first discovered by Schonbein in the 1830s. An electrochemical "cell" converts the source fuel into an electrical current and water. The process is based on reactions between a fuel and an oxidant, which is triggered by the presence of an electrolyte. The reactants flow in and the reaction products flow out, while the electrolyte remains within. They can operate virtually continuously as long as the necessary flows occur. They are not batteries.

There are many combinations of fuels and oxidants. For example, hydrogen can serve as the fuel and oxygen as the oxidant in a hydrogen fuel cell. Other oxidants include chlorine and chlorine dioxide. Other common fuels are hydrocarbons and alcohols.

The efficiency of a fuel cell is dependent on the amount of power (current) drawn from it. It seems counterintuitive, but it is true that the more that is drawn, the lower the efficiency.

Historically, fuel cells have proven very useful as power sources in remote locations, such as weather stations, spacecraft, rural areas and some military applications.

An often voiced need has been for smaller fuel cells. Bloom Energy captured the clean energy headlines at the end of the first decade of this century with its "Bloom Box."

Questions emerged as to how the Bloom Box would stack up with alternative fuels, such as solar. *Energy Business Daily* in early 2010 reported that Greentech Media had conducted a comparison study of up-front costs, versatility, energy costs, maintenance, carbon emissions and availability. The smallest Bloom server array is 100-kilowatts, which currently costs $700,000 to $800,000, so at present it is too large to serve a residence or a small commercial establishment. The Greentech Media study, therefore, found that the initial cost, versatility and maintenance all favor solar. Carbon emissions from the Bloom Box are about half the average power plant, but more than solar. Of course, the striking difference between the Bloom Box and solar is its availability. Solar can only produce power during the day and the Bloom Box produces 24/7.

Familiar corporate names like FedEx, Wal-Mart, Google and eBay all have Bloom Boxes serving some of their energy supply needs. Promoters are predicting that there will soon be Bloom Boxes small enough to serve smaller facilities and residences at a cost of roughly $2,000.

There are growing numbers of cars and buses powered with fuel cells, but their prevalence is limited by the stations available to serve them.

Geothermal

Geothermal energy is generated from the Earth's core where temperatures are reported to be hotter than the surface of the sun. The Earth is made up of a number of different layers, including a solid iron core and an outer core of very hot melted rock called magma. Since the Earth's crust is broken into plates, the magma comes close to the surface near the edges of these plates.

These naturally occurring areas are called geothermal reservoirs. Most are deep underground with no visible indication of their presence above ground. The most common areas for geothermal reservoirs are in the "Ring of Fire," which encircles the Pacific Ocean. When the geothermal reservoirs are located within a mile or two of the surface, this source is used for electricity generating power plants, geothermal heat pumps, district heating systems and direct use. This energy source is generally energy efficient, cost effective and has virtually no negative impact on the environment. The problem, of course, is that its use is constrained by the geographical location of the geothermal reservoirs.

Oregon has a long history of making effective use of its geothermal energy potential to heat buildings, as well as snow melt systems for sidewalks, stairs and handicap ramps. The Oregon Institute of Technology in Klamath Falls is well on its way to becoming not only geothermally heated but also geothermally powered. The Institute's geothermal water is produced from three wells drilled along a fault line in the southeast corner of its campus. Well water temperature varies between 192 and 196 degrees F. Typically only one well is currently used with two required during extreme cold weather. The third is used for standby, which permits maintenance without an interruption of service. [4-1]

Biofuels

The potential for biofuels dominates much of the renewable energy literature. The reason is simple: biofuels are poised to meet a major portion of the world's increasing energy demand.

At the rate we are going, transportation is destined to come to a standstill if we continue to rely solely on petroleum for transport. By 2025, experts predict that transportation fuel consumption in the U.S. will be up 25 percent and worldwide demand will be up 35 percent.

Concerns related to these demands are compounded by the apprehension of unexpected events, such as unstable political conditions in the Middle East and hurricanes in the Gulf Coast. Repercussions from the BP Deepwater Horizon disaster are still unfolding. Increasing demand plus reliability problems could bring transportation needs into critical focus within twenty years. Understandably, an alternative domestic source has increasing appeal. Biofuels could fill this need.

Biofuels are oxygenated organic compounds—methyl or ethyl esters—derived from a variety of renewable sources such as vegetable oil, animal fat, and cooking oil. More specifically, biofuels can come from human and industrial waste, such as forest/lumber, household waste and wood. Thus, biofuels also satisfy some of our growing concerns about waste disposal. As with most renewables, emissions from its use are much lower than petroleum counterparts.

An added advantage offered by biofuels is that they can be used without modification in engines and equipment. Further, biofuels are biodegradable, basically free of sulfur and considered non-toxic.

Reports coming out of the EU do not give biofuels such a clean bill of health, however, when the source of the fuel, the biomass, is held up to scrutiny. The greenhouse gases that are released converting land to corn, for example, referred to as indirect land-use change (ILUC) call into question the net environmental benefit. With reference to corn-based biofuels the U.S. EPA ruled in February of 2010 that manufacturers would need to use "advanced efficient technologies" during production if the process is to stay within U.S. limits.

Biofuels can, and are, being blended with petroleum products. Ethanol and biodiesel are the most commonly recognized biofuels to meet the transportation fuel gap. At present, the U.S. does not have a shortage of biodiesel fuel. In 2008, U.S. production of biodiesel was 16 million barrels and consumption was 7.6 million barrels. The U.S. actually exported 8.4 million barrels that year. It has become a growth industry with worldwide production more than doubling in the last five years. Government intervention has provided incentives, such as subsidies, import tariffs, consumption mandates and preferential taxation, which have stimulated production and consumption.

Biofuels, however, offer a perfect illustration of why various measures cannot be analyzed as to their benefits in isolation. The incentives of the U.S. government to produce corn-based ethanol essentially overlooked what this new demand would do to the price and availability of corn. One unintended consequence was the distortion of the food market, which took corn tortillas off many tables in Mexico as they became unaffordable. Another issue has been the sudden demand for water in areas where water availability is a serious concern. In the midwestern part of the U.S., local folks have been unhappy and suits have been filed.

There are, however, many sources of biofuel other than corn. As a result of an involved photosynthetic process, biodiesel and ethanol developers have uncovered an unlikely source: algae. Under optimal conditions, algae can double in volume overnight. Even better, up to 50 percent of algae's body weight is comprised of oil, which compares favorably with other sources, such as palm trees which yield only about 20 percent of their weight in oil. According to an *Energy Business* 2009 report, "Biofuels from Algae Market Potential": "Soy produces some 50 gallons of oil per acre per year; canola (rapeseed oil), 150 gallons; and palm, 650 gallons. But algae are expected to produce 10,000 gallons per acre per year, and eventually even more."

Algae, of course, is not without its problems, or the world's biodiesel would already be heavily supplied from microalgae. The potential is great, so the problems will very likely be resolved. Advice to CSOs? Stay tuned.

Biofuels do have a storage problem in that the oxygen contained in biodiesel makes it unstable. This means that some stabilization is required to avoid problems associated with storage. Its corrosive properties also create many transport and storage problems.

If the best available biomass conversion technology is implemented, predictions are that biofuels could replace 30 percent of the petroleum currently used for transportation.

Marine Energy

With 71 percent of the Earth's surface covered with water, there is a huge potential for energy resources. Much of that water is constantly in motion—motion that might be milked for energy.

Research in refining old technologies and creating new ones is ongoing. These include wave power generation, tidal (current) stream technologies, salinity gradient power generation, and thermal gradient.

Each of these areas has a raft of energy capture ideas. For example, in wave power generation, work goes forward in wave capture devices, shoreline devices, oscillating water columns, offshore wave energy converters, floats, wave pumps, etc. Exploring salinity gradient includes osmic power, hydrocratic generation, vapor compression and reverse electrodialysis.

As we pursue these opportunities, we must be mindful, however, that roughly 90 percent of life on Earth is in the water. New species, especially those found at great depths, are being discovered every year. It is critical that the pursuit of marine energy be done in harmony with marine life. Small mistakes could cause serious damage as well as furthering resistance to additional important research. Perceived problems can also create obstacles. At one time, there was strong resistance to the Alcan pipeline. It was most intriguing years later to see pictures in our offices at the U.S. DOE, of caribou leaning against the pipe for warmth. Fear of

WAVE POWER GENERATION

oil rig structures in the Gulf have been replaced with pictures of fish finding new homes around those structures. In fact, the rigs-to-reefs program has shown that toppled rigs make great new homes for fish.

A concerted effort to create energy sources in harmony with marine life could prove to be extremely valuable, but a solid communications effort must accompany the effort. Planned strategies to inform the public of the compatibility must go hand in hand with the developing technology.

Hydropower

Closely aligned with marine energy is the hydropower gained from major rivers. Energy derived from moving water now contributes about 20 percent of the world's electricity production. There is about 650,000 MW installed and approximately 135,000 MW under construction, or in the final planning stages.

If we reach back in history we can find examples of hydropower used to run saw mills, mill grain and manufacturer textiles. The control of water by dams also provided for new arable land for farming.

Hydropower is a very attractive source of power because it is renewable and produces essentially no harmful emissions into the atmosphere. Hydropower is also less expensive than electricity generated from nuclear or fossil fuel sources.

There is a move afoot to demolish the dams. Proponents of this effort view believe it is a step toward returning the Earth to its natural state. Careful examination of such proposed efforts shows that in many cases it is not feasible. Most of the discourse is about river flow, natural vs. man-made, and does not address the energy implications.

There are large untapped sources of hydropower throughout the world. For areas expected to have great growth in power demand in the years ahead, this is a very economically viable and environmentally benign source. Careful consideration needs to be given to developing more hydropower from government and socially responsible equity funds. Hydropower is not viewed as a major source of new power in industrialized nations as the available power has been exploited, or is considered unavailable for environmental reasons.

In examining the potential which hydropower offers, many factors need to be weighed. Maximum exploitation of some sites may have a negative impact on land use. These losses need to be assessed in relation to the clear benefits hydropower offers in comparison to other electric power generation sources.

NON-RENEWABLES

Nuclear Energy

Just as renewables are not necessarily low emission sources, non-renewables are not necessarily high emission sources. Nuclear power is an excellent example of a non-renewable that is a low emission source.

It is almost impossible to view the potential for nuclear energy objectively. Any consideration is clouded with concerns for safety and the association of this power source with nuclear armaments. A few observations may help.

First, this source of power should properly be called uranium, not nuclear. All of the uranium used in the U.S. generates electricity. It represents 19.4 percent of the electricity generated in the states.

Since new construction of nuclear plants has been quiet for so long in the U.S., it often comes as a surprise to Americans that the U.S. has, at 30 percent, the greatest share of nuclear electric generation in the world. Other major shares of the world's nuclear power belong to France at 16 percent, Japan at 11 percent, Germany at 6 percent and Russia and South Korea at 5 percent each. Of the 443 operable reactors in the world, 104 of them are in the U.S., and they contribute 8 percent of our total energy supply. Thirty-one states have commercial nuclear plants with the highest ranked being Illinois with 6 and Pennsylvania with 5. The best known is probably Three Mile Island in Pennsylvania because of an incident there many years ago. The story *not told* about Three Mile Island, however, is that when problems occurred, the back-up system worked.

Nuclear energy gets its name from the energy generated from the nucleus of an atom. This can be achieved by splitting the atom through fission or when atoms are combined through fusion. A present, fusion is not commercially viable.

The atom that is split is uranium 235. Uranium can be found in rocks around the world, but uranium 235 (U235) is relatively rare. The U.S. produces, domestically, about 14 percent of U235 that it uses. Of the 86 percent imported, most of it (42 percent) comes from Australia and Canada. Roughly 33 percent is imported from

Kazakhstan, Russia and Uzbekistan. The remainder comes from Brazil, Czech Republic, Namibia, Niger, South Africa and the UK.

The biggest problem associated with nuclear energy is the storage of spent fuel. The highly radioactive properties are a serious concern. A huge storage facility was under construction in Nevada until the federal government, without explanation, stopped work in 2010. This has created uncertainty, which in turn has slowed the recently renewed efforts to construct more nuclear plants in the U.S.

As indicated in Chapter 3, a major sustainability issue is water. A major international effort under the International Atomic Energy Agency to create economically viable designs for nuclear desalination plants is under way.

From Rio to Kyoto to Copenhagen, there has been a strong, concerted effort, to reduce emissions by country. At play is the position of the Peoples Republic of China (PRC) *vis-à-vis* the U.S. The PRC at present has a very aggressive campaign to build nuclear plants. One of the people closely associated with this endeavor believes a major part of the agenda is to place China in a stronger position, through these nuclear energy efforts, to challenge the U.S. to reduce its carbon footprint.

Coal

The most abundant fuel produced in the U.S. is coal. It is actually a sedimentary rock composed mostly of carbon and hydrocarbons. It has taken millions of years for coal to form from plants that lived hundreds of millions of years ago.

There are four main types, or ranks, of coal: anthracite, bituminous, sub-bituminous and lignite. The classifications depends on the amount and type of carbon the rock contains and the amount of heat energy it can produce. Anthracite contains 86-97 percent carbon with a heating value slightly higher than bituminous, but accounts for less than 0.5 percent of the coal mined in the U.S. Formed under high heat and pressure, bituminous coal is 45-86 percent carbon and is valued for generating electricity and as a raw material for steel and iron. Sub-bituminous coal, which represents about 45 percent of the coal produced in the U.S., has a lower heating value that bituminous and usually is about 35-45 percent carbon. The lowest ranked coal is lignite with only 25-35 percent carbon and constitutes about 7 percent of US coal.

Coal is mined from the ground with about two-thirds coming from coal within 200 feet of the ground's surface. Modern mining methods have allowed us to triple the amount of coal produced since 1978 as measured by one miner/hour. Coal is mined in 26 states with Wyoming, West Virginia, Kentucky, Pennsylvania and Texas offering the most coal. Some of the U.S.'s cleanest burning coal was taken off the market in the 1990s when President Clinton designated the land where it was found as a federal preserve. The dominant source of this cleaner burning coal is now Indonesia.

Large amounts of coal are used in the concrete and paper industry. Separated ingredients of coal, such as methanol and ethylene, are used to make plastics, synthetic fibers, medicines, tar and fertilizers. The majority of coal used in the U.S. (93 percent) is used to generate electricity.

Environmental laws and modern technology have greatly reduced the environmental impact of the production and consumption of coal. Carbon capture in coal-fired power plants has helped reduce the emissions and more such plants are being built. Figure 4-1 shows the major components in such a plant.

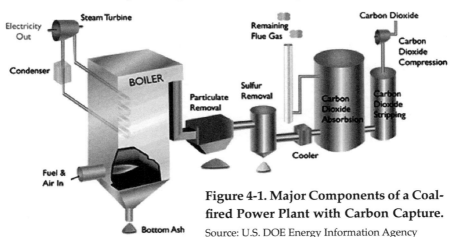

Figure 4-1. Major Components of a Coal-fired Power Plant with Carbon Capture.

Source: U.S. DOE Energy Information Agency

Bio-coal

Our range of options continue to grow. One of the more recent advents has been the development of bio-coal. This process changes the chemistry and structure of wood into friable material with high carbon content. Bio-coal can be used in conjunction with coal or to replace it. Bio-coal is also available in briquettes, which offer a high density and energy concentrated fuel, in a size that can be more manageable. The economic potential could be huge, but at this point we are mostly in uncharted waters.

UNCHARTED WATERS

Oil

The dominant non-renewable fuel sources is, of course, oil. At 19.5 barrels per day, the U.S. is the top oil consuming country in the world. Over 60 percent of the oil consumed in the states is currently imported. Of that consumption, roughly 70 percent is used in the transport sector.

Oil is divided by various labels. The amount of sulfur in the oil determines whether it is "sweet" or "sour." The weight of its molecules classifies it as "light," which flows like water, or "heavy," which is the consistency of tar.

The amount of oil in our earth is considered finite as it takes millions of years for the remains of plants and animals to be converted by heat and pressure into oil. The term "petroleum" means "oil from the earth." The top crude oil-producing countries of the world are Russia, Saudi Arabia, U.S., Iran and China. About one-fourth of the crude oil produced comes from the Gulf of Mexico. The BP disaster in 2010 made us all very aware of the potential marine/shore fragility. The moratorium placed on drilling had a huge impact on the Gulf states' economies. The longer range impact has yet to be fully assessed. The newer, more technologically advanced, safe drilling rigs were relocated in other countries. These rigs represent major investments and cannot economically remain idle. Even though the moratorium prompted the federal government to make a stronger push for renewable sources, the net result has been an increase in U.S. dependence on foreign oil. The extent has yet to be realized.

The world's oil and gas industry is dominated by national oil companies. When it comes to "big oil," many point to Exxon Mobil, but in terms of reserves, it is ranked 14th behind companies owned by other governments, many of which are not friendly to the U.S. The U.S. can take some solace in the fact that its current number one supplier of petroleum is Canada at 2.5 million barrels/day.

"Upstream" activities, which involve exploration, development and production of crude oil and natural gas, are particularly active in the Asia-Pacific theatre. "Midstream" activities, which include transportation of oil and gas through pipelines and tanks, are of growing interest throughout the world. For example, the new natural gas pipeline China has opened to central Asia could alter the geopolitics of the world.

Crude is moved by pipeline, ship or barge to refineries. As shown in Figure 4-2, a barrel of crude oil typically yields18.56 gallons of gasoline, 10.31 gallons of diesel, 4.07 gallons of jet fuel, less than 2 gallons each of liquefied petroleum gas, heavy (residual) fuel oil, and heating oil. About 7 gallons of other products are also produced.

Products Made from a Barrel of Crude Oil (Gallons) (2008)

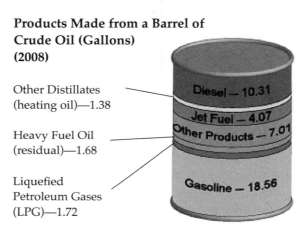

Other Distillates (heating oil)—1.38

Heavy Fuel Oil (residual)—1.68

Liquefied Petroleum Gases (LPG)—1.72

Diesel — 10.31
Jet Fuel — 4.07
Other Products — 7.01
Gasoline — 18.56

Figure 4-2

Of course, the value of these products comes from burning them to yield energy. While petroleum products make our life easier and have become essential to our current way of life, the finding, producing, moving and burning also produce some undesirable by-products which pollute the air and water, harming our environment. The most common emissions are carbon dioxide (CO_2), carbon monoxide (CO), sulfur dioxide (SO_2), nitrogen oxide (NO_x) and volatile organic compounds (VOCs). Also of concern is the particulate matter (PM) which causes the hazy conditions in cities. Along with ozone, PM creates health problems, such as asthma and chronic bronchitis.

Laws and technology have done much to reduce the negative impact of burning fossil fuels. "Reformulated fuels" used for transport are much cleaner than those used twenty years ago. Exploring and drilling technology improvements have reduced the "footprint" through the use of satellites, GPS, remote sensors and seismic technologies. Horizontal and directional drilling have vastly improved underground access from a single surface location.

When a well is no longer economically viable, it is plugged underground and becomes virtually undetectable. In the "rigs and reefs" program, old offshore rigs are toppled and rapidly become habitats for marine life.

While oil spills from tankers are well-covered in the press and can harm wildlife, the major problem of oil in marine waters comes from naturally occurring leaks from the ocean floor. This seepage contributes about 92 percent of the oil in ocean water. Of course, the numbers and the percentages changed dramatically in 2010 with the Deepwater Horizon disaster. Contamination projections have also been thrown askew by the federal government's moratorium on off-shore drilling.

Many advances have been made to reduce oil in water, including double hulls on new ships and double lining on underground storage tanks. There are other ways, however, in which humans treat fuels that create a significant impact. The dripping of gasoline when filling our gas tanks and way we dispose of motor oil are examples of the way individual behavior can contribute to the oil in gutters to rivers to oceans.

As long as oil and gas remain the dominant energy sources in the world, the political and economic ramifications have the potential to be huge. We have only to recollect the long gas lines at the stations, the order for lights out at all monuments, etc. during the 1973 Oil Embargo, and to remind ourselves that we were only importing 26 percent at the time. Envision what the U.S. economy and our way of life could be today, with our heavy dependence on foreign oil, should that foreign supply be stopped.

Natural Gas

Natural gas is an interesting phenomena in our energy repertoire as it is a colorless, odorless and tasteless fuel. In fact, mercaptan (a chemical that smells like sulfur) is added before distribution as a safety device so we can detect leaks.

Natural gas is considered a clean fuel compared to other fossil fuels, as it has fewer emissions of sulfur, carbon, and nitrogen. Burning natural gas, does produce carbon dioxide, which is a greenhouse gas, and contributes to our environmental problems.

Pressure, time and heat changed the remains of plants and animals into an organic material trapped beneath the earth's surface. The main ingredient in natural gas is methane, a compound of one carbon atom and four hydrogen atoms. Natural gas is found by conducting seismic survey using echoes from vibration source at the Earth's surface, or from studying rock samples. If a site seems promising, drilling begins. Some of these sites are on land and many are offshore.

Most of the natural gas consumed in the U.S. is produced domestically. Some is imported from Canada.

We also use machines called "digesters' that turn today's organic material (animal wastes, plants, etc.) into natural gas. In effect, this process replicates the natural processes in the Earth, but shortens the waiting period by millions of years.

One of the most pleasant surprises in the energy community in recent years is the increased accessibility of natural gas.

Typically, natural gas is moved by pipelines from the producing fields to consumers. There is often resistance to these pipelines being laid in certain areas.

In addition to heating, natural gas is used to produce steel, glass, paper, clothing, brick and electricity. It is also a raw material for many common products including plastics, paints, fertilizer, medicines and explosives. As shown in Figure 4-3, 27 percent of the natural gas consumed in the U.S. in 2009 was used for industrial purposes, 30 percent was used for electric power generation, 21 percent was residential and 14 percent was used for commercial purposes.

LNG

Increasingly natural gas is also being shipped to the U.S. as liquefied natural gas (LNG). To do so, it is chilled to very cold tem-

Natural Gas Use, 2009

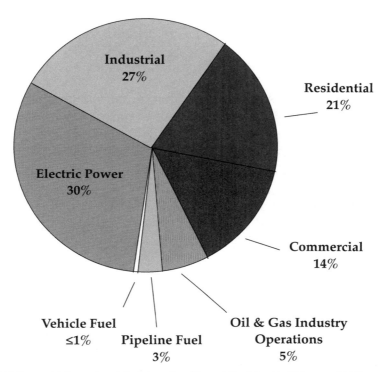

Source: U.S. Energy Information Administration, *Natural Gas Monthly* (February 2010).

Figure 4-3. Natural Gas Use, 2009

peratures, about –260°F. At this temperature, it changes into a liquid and can be stored in this form. It has the added attribute of taking up only 1/600th of the space that it would need in its gaseous state.

The global LNG market is growing rapidly. This year's output in millions of tonnes per annum (MMTPA) is expected to double the flow from 2000 when it was 113 MMTPA. By 2030, it is expected to double again to roughly 524 MMTPA.

Today, 33 countries are involved in global LNG. The combined liquefaction capacity is 402 MMTPA from 19 countries.

Prospects outweigh challenges and growth is expected to continue in all regions of the world.

ENERGY EFFICIENCY: ZERO GENERATION

"The greenest electron is the one not generated." A school administrator from West Alice, Wisconsin, stated this during an online seminar I was giving several years ago. Variations of this statement have come to my attention over the years, but this voiced observation for me carries the greatest impact. It says it all.

Energy efficiency allows us to use less to achieve the same purpose. Energy efficiency means less needs to be generated to do specific work. When electrons are NOT generated, no contaminates are emitted.

In our effort to reduce pollution, we sometime overlook the fact that electrons not generated also represent an opportunity to conserve our finite resources. In our zealousness to find alternatives to oil we have tended to categorize oil as all bad. To weigh all our supply options fairly, we must recognize the unprecedented role oil has played in the evolution of our world. Certainly, the use of oil presents us with environmental problems, but it has also served some valuable purposes. Between now and the time alternatives will meet our energy needs, oil will continue to serve a critical, valued purpose. Energy efficiency can help preserve our resources to meet these needs.

Achieving greater energy efficiency and its tremendous value in meeting our future energy demands was discussed in the previous chapter. Here EE is judged as a supply option. As such, it is well to remember that EE can produce revenue. While all other sources of energy will cost, EE can pay for itself and even generate a positive cash flow.

EE preserves our precious resources (a cornerstone of sustainability) and generates revenue while avoiding any pollution. In summary, it is the best source of energy available. Wherever it is economically viable to use EE, it should be the preferred option in any sustainability program.

Closely allied with energy efficiency is the opportunity to use cogeneration, often referred to as combined heat and power (CHP). The great value of CHP is the ability to gain both heat and power from one energy source. There is also the advantage of being able to generate electricity on site, which reduces dependence on our troubled grid and increases the reliability and availability of electricity.

WEIGHING THE OPTIONS

Renewable energy is an increasingly attractive source of power. Government incentives in many parts of the world are fostering more reliance on renewables, and by 2030, our reliance on renewables is expected to double.

The growth of renewables is encouraging. At 7,315,711 billion Btu, the use of renewable energy in 2008 was 7.5 percent greater than the preceding year. Figures for that year show that hydroelectric power accounted for 34 percent and biomass for 53 percent of the total renewable energy consumption. Contributions from other renewables were 7 percent from wind, 5 percent from geothermal and 1.23 percent from solar. [4-2]

Realistically, however, renewables will remain relatively costly in the near term and will be slow to fill the energy demand gap. For the next two decades, the major reliance will continue to be on fossil fuels. The U.S. DOE Energy Information Agency projects the use of coal will grow faster than liquid fuels and natural gas.

Despite our concerns related to the use of fossil fuels, it is essential to our getting from *here* to *there*. We must constantly look for ways to use alternative energy and to weigh the consequences of our actions. Then, the only remaining question is: will the U.S. continue on its path of energy dependence on foreign sources. It's worth repeating: it is NOT a question of renewable vs. fossil fuels;

it is a question of OUR oil vs. THEIR oil.

If we are faithful to our sustainability management goals in preserving our planet *in its entirety;* then, the concern becomes one of who can explore, drill and transport the oil with the least harm to our Earth.

Decisions of immense consequences are made globally, nationally, regionally *and locally.* CSOs must become knowledgeable regarding the facts and their implications, weigh the options carefully, and act accordingly.

Chapter 5

Sustainable Buildings

Stephen Roosa

A design that is green is one that is aware of and respects nature and the natural order of things; it is a design that minimizes the negative human impacts on the natural surroundings, materials, resources, and processes that prevail in nature. It is not necessarily a concept that denies the need for any human impact, for human existence is part of nature too. Rather, it endorses the belief that humankind can exist, multiply, build and prosper in accord with nature and the earth's natural processes without inflicting irreversible damage to those processes and the long-term habitability of the planet.

GRUMMAN

Buildings and structures play an important role in meeting human needs. They provide shelter and can provide healthy environments. Buildings, however, can also provide unhealthy conditions and stress the environments in which they are built. Buildings require excavation for construction, can modify wildlife habitats, change rainwater runoff patterns, change landscapes, and require a cornucopia of materials. They also absorb and radiate heat, need paving for pedestrian and vehicles and, most importantly, consume resources—including energy. Energy consumption varies with building size, design and climate conditions. Chicago's Sears Tower alone consumes more electricity than Rockford, Illinois, a city with a population over 150,000.

In the U.S., buildings represent 39 percent of primary energy use, 70 percent of electrical consumption, 12 percent of potable water use, and generate 136 million tons of construction waste annually.[5-1] If there were a way to reduce electrical demand in buildings by only 10 percent, most of the new electrical generating facilities scheduled for construction in the U.S. over the next 10 years would

91

not be required. The carbon associated with the energy used in U.S. buildings "constitutes 8 percent of the current global emissions—equal to the total emissions of Japan and the United Kingdom combined."

Buildings have an economic life span. Many are often financed for 30 years or more and have been designed to serve for 40-50 years. Over time, buildings are subject to deterioration, exposure to the elements, changes in use and occupancy, changes in building standards, and changes in ownership. With this in mind the idea of a 'sustainable building' may seem to be a contradiction. However, buildings and building materials can often be recycled and reused. Older structures can be updated to serve new purposes, recycling their existing infrastructure. Perhaps the ultimate recycling project is moving a building to another location rather than demolishing it. In such cases, a substantial portion of the energy and material embedded in its original construction can be reused.

Figure 5-1. Moving a Building—The Ultimate Recycling Project

The idea of developing sustainable buildings stands in striking contrast to typical standard construction practices. Sustainable construction involves changes in the fundamental design processes, refocusing on-site development, rethinking materials and building components, considering environmental impacts, and determining how healthy interior environments can be configured. To further complicate the process of designing and developing green buildings, Huber concludes that "…building sustainability strategies have to consider basic societal conditions. Building regulations, financing models, rental legislation, environmental legislation, etc., significantly determine the construction design. The solution is approached by an eco-efficient optimization which considers the basic societal conditions."

Knowing that buildings require substantial inputs of resources to construct and maintain, what can be done to reduce their environmental impacts and energy requirements? The solutions include, but are not limited to:

1) Changing land development practices; 2) designing buildings with attention to improved construction standards; 3) upgrading and reusing existing structures; 4) providing more efficient buildings and higher quality equipment; 5) changing the physical arrangement and configuration of buildings; 6) carefully selecting construction materials; and 7) harvesting on-site resources.

This chapter considers how buildings can employ green practices and how these technologies are incorporated into the design of buildings. "Green building programs" will also be discussed. These programs include efforts to establish energy and environmental standards for existing buildings and new construction.

LAND DEVELOPMENT PRACTICES

Land development practices have contributed to adverse environmental consequences and changes in infrastructure that have increased energy and water consumption. Large-scale developments such as manufacturing centers and residential subdivisions constitute a type of "terraforming," which cause permanent changes to their environs. The growing appetite of these structures for energy and environmental resources has contributed to ecosystem disruption and has forced us to rethink how buildings are sited and constructed.

While development is associated with localized environmental disruptions, scientific assessments of construction impacts have yielded mixed results. We know that construction is energy intensive and resources are consumed in the process of construction. Often, initial costs are minimized, and life-cycle cost analysis is not used in the selection of materials and mechanical systems. When initial costs are minimized, maintenance, energy, and water costs typically increase. Development that favors sub-urbanization and automobile-biased transportation systems also increases resource costs. Green development practices coupled with design alternatives for structures with environmentally friendly and energy efficient attributes have now become a feasible alternative. When these are addressed in the site selection and design of buildings, life-cycle costs are typically reduced.

Sustainable development focuses on the built environment: buildings are seen as the primary building blocks of the urban infrastructure. Green construction practices provide alternatives. It is understood that "being green" implies a commitment to environmental protection and natural resource conservation. If buildings can be constructed in a manner that is less environmentally damaging and more energy efficient, then they can be called "green." Being green means that construction needs to be "eco-efficient." Life-cycle cost assessments should be made in selecting building systems and construction components. As DeSimone and Popoff noted, "Eco-efficiency is achieved by the delivery of competitively priced goods and services that satisfy human needs and quality of life, while progressively reducing ecological impacts and resource intensity throughout the life-cycle to a level at least in line with the earth's carrying capacity."

In regard to the built environment, architectural designers

have recently renewed their emphasis on designing healthy buildings. Fundamental design issues such as site orientation, daylighting, shading, landscaping, more thermally cohesive building shells, and more energy efficient mechanical and electrical systems, are getting renewed attention.

Drivers include experiences with indoor air quality issues and illnesses linked to systemic malfunctions such as chemical exposures, legionnaire's disease and asthma. Medical conditions have been linked to mold exposures that result from indoor environmental conditions. These concerns served to jolt architects and engineers into re-establishing the importance of indoor environmental conditions in general, and indoor air quality in particular, when designing their buildings. With concerns mounting as to product safety and liability issues involving the chemical composition of materials, manufacturers have begun to mitigate the potential adverse impacts of building materials upon occupants. Concurrently, resource availability and waste reduction became issues for both building design and the design of construction components.

Central to the chief sustainability officer's responsibilities are decisions as to what types of buildings to construct, what construction standards should apply, and what sorts of materials should be to used in the construction of buildings. Supporting the chief sustainability officer (CSO) are those who directly influence decisions regarding the physical form of a proposed structure including the builders, developers, contractors, architects, engineers, planners and local zoning agencies. With so many parties involved in the decision making process, planning for green buildings can be a difficult, decentralized and often divisive process. All involved must abide by regulations that apply to the site and structure being planned. Planning ordinances and building codes often vary from one locale to another, further complicating the process. For these reasons, making green buildings requires a team of cooperative individuals. It is a comprehensive process that involves careful design and selection of building components and systems.

What is alarming is that past professional practice within the U.S. building industry has rarely gauged the environmental or energy impact of a structure prior to its construction. Building types, especially for commercial construction, are often standardized, regardless of differentiators, such as local climate conditions, geography and site conditions. This lack of differentiation contributes to increased resource consumption. Green buildings are an attempt to improve the planning and design of new structures and major renovations.

THE CONCEPT OF GREEN BUILDINGS

There was a time in the U.S. when most construction material was obtained locally. Indigenous materials included accessible timber, fieldstone, locally quarried rock, adobe, thatch, slate, clapboard, and cedar shakes. Since construction materials were costly to manufacture (most were hand-made) and troublesome to transport, most components of demolished structures were reused in some manner.

Early in U.S. history, one- and two-room log houses were the norm. The central heating system was a drafty fireplace with a chimney constructed of local stone. When possible, design features, learned by trial and error, were added in an attempt to optimize thermal comfort. Examples included architectural features. which would control lighting and temperature in the indoor environment with shading devices, orientation to breezes, and designs to carefully size and orient fenestration. Rainwater was often collected from roofs. Such were the humble beginnings of green construction practices.

Buildings today are infinitely more complex—and sustainable building practices can be even more so. Accepting the notion that sustainable, environmentally appropriate, and energy-efficient buildings can be labeled "green," the degree of "greenness" is subject to multiple interpretations. The process of determining which attributes of a structure can be considered "green" or "not green" can be viewed as inconclusive and subjective.

A green design solution appropriate for one locale may be inappropriate for others, due to variations in climate or geography.

Figure 5-2. LEED Certified Building—Lincoln Hall, Berea College

Complicating the process, there are no clearly labeled "red" edifices with diametrically opposing attributes. While it is implied that a green building may be an improvement over current construction practice, comparison is often unclear and confusing. It can often be perplexing as to what sort of changes in construction practice, if imposed, would lead to greener, more sustainable buildings.

At times, markets adjust and provide materials, components and products so that greener buildings can arise. Since standards are often formative and evolving, gauging the degree of "greenness" risks the need to quantify often subjective concepts.

One attribute of green construction practices is the attempt to preserve and restore habitat that is vital for life, or to become "a net producer and exporter of resources, materials, energy and water rather than being a net consumer."[5-2] As described in the Governor's Green Government Council, the Pennsylvania Council defines a green building as "one whose construction and lifetime of operation assures the healthiest possible environment while representing the most efficient and least disruptive use of land, water, energy and resources."[5-3]

There are qualities of structures, such as reduced environmental impact, improved indoor air quality and comparatively lower energy usage, which are widely accepted as evidence of green construction practices. For example, using recycled materials that originate from a previous use in the consumer market, or using post-industrial content that would otherwise be diverted to landfills, are both widely accepted green construction practices.

Grumman describes a green building by saying that "...a green building is one that achieves high performance over the full life cycle." Performance can be defined in various ways. "High performance" can be interpreted widely, and often in "highly" subjective ways. In an attempt to clarify, Grumman further identifies a number of attributes of green buildings:

•   Minimal consumption—due to reduction of need and more efficient utilization—of nonrenewable natural resources, depletable energy resources, land, water, and other materials.

•   Minimal atmospheric emissions having negative environmental impacts, especially those related to greenhouse gases, climate change particulates, or acid rain.

•   Minimal discharge of harmful liquid effluents and solid wastes, including those resulting from the ultimate demolition of the building itself at the end of its useful life.

•   Minimal negative impacts on site ecosystems.

•   Maximum quality of indoor environment, including air quality, thermal regime, illumination, acoustics, noise and visual aspects.

While goals and attributes of green construction practices are readily identified, developing construction standards to achieve such goals is another matter. High performance sounds like added value, for which a premium is likely to be paid. In fact, adding green building features is likely to increase construction costs by 1 to 5 percent, while lowering operating expenses and life-cycle costs. Energy-efficient electric motors may be slightly more expensive to install but the savings can be substantial. An estimated "97 percent of the life-cycle costs of a standard motor goes to energy costs and only 3 percent to procurement and installation."[5-4].

Buildings are further differentiated by the desires of the owners, the skills and creativity of their design teams, site locations, local planning and construction standards, and a host of other conditions. What may be a green solution for one building might be inappropriate if applied to another.

There are a multitude of shades of green in green building construction practices. The qualities of green construction practices have been variously identified.[5-5] Some qualities focus on exterior features and others on the types of green materials that are used. The International Energy Conservation Code (IECC), for example, requires that energy efficient design be used in construction and provides effective methodologies. Its focus is on the design of energy-efficient building exteriors, mechanical systems, lighting systems and internal power systems. The IECC is being included in building codes across the U.S. and in many other countries.

ASHRAE Standard 90.1 (Energy Standard for Buildings Except Low-Rise Residential Buildings) deals with the energy-efficiency of buildings and HVAC components. Other standards concern improved air quality and better ventilation systems. Green construction opportunities are further described in the ASHRAE Green Guide. Green buildings may also be designed to focus carbon impact and some are classified as "zero-carbon," meaning that they have a net zero carbon emissions impact.

COMPARATIVE BUILDING ENERGY PERFORMANCE

Since new construction practices are often tested in housing applications, residential construction provides heuristic examples.

Making buildings stingy in their use of energy is one aspect of green construction. There are many opportunities to improve energy performance in residential, industrial, and commercial buildings. Advances in construction practices have yielded striking results and provide the means to create more efficient buildings.

The chapter author conducted a one-month (December 2006) heating and electrical system assessment of two occupied residences in the mid-western U.S. during the same time period. The homes were about 16 km (10 miles) apart. Only energy consumption was considered in this assessment. Both residences were two-story buildings, located in the same metropolitan area, served by the same utility companies, and on the same rate structures for electricity and natural gas. Neither used alternative energy sources. For the period of comparison, both residences were heated with natural gas with thermostats set at approximately 22°C (72°F) and neither used temperature setback controls. Both used high-efficiency (approximately 90-94 percent) natural gas fired, forced-air furnaces to heat interior spaces.

One building was a frame residence constructed in 1910 that was approximately 212 m^2 (2,280 ft^2). This frame residence had limited insulation, single pane windows, electric water heating and used primarily incandescent lighting. The other residence (designed by the author) was a brick home in a suburban location constructed in 2001. Both residences used the conventional construction practices available at the times of their construction. The newer 427 m^2 (4,600 ft^2) brick residence was designed with an expanded south-facing façade, reduced northern exposure, extra insulation in walls, an exterior infiltration wrap, insulated ceilings and foundation, high efficiency windows, natural gas water heating, extensive use of compact fluorescent lighting, and other features. To take advantage of topography and the thermal moderation available from the earth, it was set into the slope of a hill, minimizing the northwestern exposure and maximizing the solar gain from the southwest. The total energy bills for the older 212 m^2 frame structure for the period of study totaled $432, equating to $2.24/m^2 (19 cents/ft^2). The total energy bills for the newer 427

m^2 brick residence for the period totaled \$275, equating to \$.58/m^2 (5.2 cents/ft^2).

This comparison provides interesting findings. For the newer brick structure, total energy costs were 36 percent less than the older frame residence for the period of study despite the fact that it was twice as large. The energy costs for the newer residence (based on a unit area) were roughly 74 percent less for the period. The older frame residence used 556 kWh of electricity and 1,001.3 m^3 (353.6 ccf) of natural gas while the newer residence used 1,182 kWh of electricity and 519.9 m^3 (183.6 ccf) of natural gas. The newer residence consumed twice as much electricity and half as much natural gas during the study period. The increased electrical use was likely due to greater areas of space for lighting. The lower natural gas usage was likely the result of the improved thermal envelope.

While not a scientific study, the analysis provides empirical evidence that energy costs can be lower when energy efficiency technologies and conventional construction practices are incorporated in the design of residences.[5-6] It also demonstrates how the consumption of source fuels in buildings of similar function can vary significantly.

ENERGY STAR BUILDINGS

In an effort to provide information to improve the energy efficiency of buildings in the U.S., agencies of the central government co-sponsored the development of the EnergyStar™ program. In 2007, the U.S. EPA described this program as providing "technical information and tools that organizations and consumers need to choose energy-efficient solutions and best management practices."[5-7] The CSO needs to be aware of the type of technical information available, including information about new building designs, green buildings, energy efficiency, networking opportunities, plus a tools and resources library. EnergyStar offers opportunities for organizations and governments to become partners in the program. The program provides guidelines to assist organizations in improving energy and financial performance and offers their partners the opportunity to be distinguished as environmental leaders. The multi-step process involves making a commitment, assessing performance, setting goals, creating an action plan, implementing the action plan, evaluating progress and recognizing achievements.[5-8]

Expanding on their success, Energy Star™ developed a building energy performance rating system, which has been used for tens of thousands of buildings. It does not claim to be a green building ranking system, but rather a comparative assessment system that focuses on energy performance, an important component of green building technologies.

GREEN CONSTRUCTION MATERIALS AND METHODS

Material and product recycling, which has a long history, and green construction have their roots in byproduct recycling. During World War II, strategic materials, such as steel and aluminum, were recycled and reused to manufacture military equipment. After the end of the war, recycling programs fell into decline. Beginning anew in the 1970s, metals such as aluminum, copper and steel began to be recycled. By the 1980s, construction site wastes, such as steel frame windows and their glass panels, were being recycled rather than being sent to landfills. By the 1990s landfill space became more costly. In addition, once-flared natural gas from landfills began to be seen as a potential energy resource rather than a waste by-product.

In 2003, Clayton defined a "sustainable" building philosophy as "to design, build and consume materials in a manner that minimizes the depletion of natural resources and optimizes the efficiency of consumption." Green construction practices have several categorical commonalities:

Green buildings are designed to reduce energy usage while optimizing the quality of indoor air. These buildings achieve energy reductions by using more insulation and improved fenestration, by optimizing the energy usage of mechanical and electrical building subsystems. They often contribute to "green" by the use of alternative energy.

There is an emphasis on reducing the costs of energy used to transport material to the construction sites. One means of achieving reductions in transportation costs is to use materials that have been locally manufactured. This provides the added benefit of supporting local employment and industries.

There is a focus on using recycled construction materials (such as reusing lumber from demolished structures) or materials made from recycled products (such as decking materials that use recycled plastics). The idea is to reduce the amount of virgin material required in the construction.

There is a preference for materials that are non-synthetic, meaning that they are produced from natural components, such as stone or wood, etc. This often reduces the number of steps required for product manufacture and may also reduce the use of non-renewable resources employed. For example, it can reduce the use of oil required in the production of plastics. Extracted metals such as aluminum and copper may also be preferred as they can be more easily reused once the building's life cycle is completed.[5-9]

There is a mandate of green construction to avoid the use of materials that either in their process, manufacture or application, are known to have environmentally deleterious effects or adversely impact heath. Examples include lead in paint or piping, mercury thermostats, and solvents or coatings that may outgas fumes and materials which may act as carcinogens.

In an effort to reduce water from municipal sources and sewage treatment requirements, green buildings are often designed to harvest rainwater by using collection systems. Rainwater may be used for irrigation, toilets or other non-potable requirements. In addition, green buildings use technologies, such as flow restriction devices, to reduce the water requirements of the building's occupants.

The quantities of construction wastes are reduced. This occurs by strategically reducing wastes generated during construction and by reusing scrap materials whenever possible. The goal

Figure 5-3. Construction Waste

is to reduce the amount of material required for construction and to reduce the quantities of scrap material that must be trucked to a landfill.

CSOs need to make procurement and general contractors aware of new industries and entire product lines that have emerged in an effort to provide construction materials that meet green building standards. There are numerous creative examples. Beaulieu Commercial has brought to market a carpet tile backing made from 85 percent percent post-consumer recycled content using recycled plastic bottles and glass, yielding carpet backing that is 50 percent stronger than conventional carpet backing. The Mohawk Group manufactures carpet cores using a recycled plastic material rather than wood. This change saves the equivalent of 68,000 trees annually. There are products on the market; e.g., ProAsh, made with fly ash—a once wasted by-product—which has been recycled into concrete. Armstrong now offers to pick up old acoustical tile from renovation projects and will deliver them to one of their manufacturing sites to be recycled into new ceiling tiles.

PVC products became available to satisfy piping needs in domestic applications for drainage systems rather than copper, lowering the weight of products and their components. The installation times are reduced by eliminating the need to sweat pipes, which lowers labor costs. While PVC has its own associated environmental issues, it is now being used for exterior trim due to its durability.

Many wood products are certified and labeled if they have used environmentally appropriate growing and harvesting techniques in their production. Pervious paving systems, that allow vegetation to grow and reduce the heat absorption from paving, are also available.

The list of available green technologies and products seems endless. Triple pane, low "e" window glazing, solar voltaic roof shingles, waterless toilets and solar powered exterior lighting systems are among the green building products that are being used in building systems.

Green construction has provided opportunities to introduce new product lines and the movement to green construction practices offers ready markets for the products. Hydrotech manufactures a green roof system that provides a balance between water drainage and soil retention thus allowing roof gardens to flourish.

To learn which products are greener than others, CSOs will find that computer software tools are available to assist in determining the environmental impact of specific products. These programs also help ascertain product life-cycles.

RATING SYSTEMS FOR GREEN BUILDINGS

While green building components are available, incorporating them into green buildings requires forethought, engineering, and creative design. Green building standards have been developed by both private and governmental organizations, all bent on finding ways to assess green construction practices.

Comparing the degree of "greenness" from one building to the next is difficult. One solution is the use of categorical rating systems in an effort to reduce subjectivity. The development of green building attributes or standards by private organizations recognizes that decisions are to be based on stakeholder consensus. These stakeholders are often from widely diverse industries and geographic locations.

Developing a rating system for green buildings is both difficult and challenging. As Boucher commented in 2004, "the value of a sustainable rating system is to condition the marketplace to balance environmental guiding principles and issues, provide a common basis to communicate performance, and to ask the right questions at the start of a project." Rating systems for sustainable buildings began to emerge in the 1990s.

Perhaps the most publicized of these rating systems first appeared in the U.K., Canada, and the U.S. In the U.K., the Build-

Figure 5-4. Pervious Paving System in Central Hungary

ing Research Establishment Environmental Assessment Method (BREEAM) was initiated in 1990. BREEAM™ certificates are awarded to developers based on an assessment of performance in regard to climate change, use of resources, impacts on human beings, ecological impact and construction management. Credits are assigned based on these and other factors. Overall ratings are assessed according to grades that range from pass to excellent.

The Canadian International Initiative for a Sustainable Built Environment (IIS-BE), based in Ottawa, Canada, has a Green Building Challenge program, which is now used by more than 15 participating countries. This collaborative venture provides an information exchange for sustainable building initiatives and has developed "environmental performance assessment systems for buildings" can be found in IISBE 2005. The IISBE has created one of the more widely used international assessment systems for green buildings.

The U.S. Green Building Council (USGBC), an independent non-profit organization, established in 1995, grew from just over 200 members in 1999 to 11,500 members by 2010. The core purpose of the USGBC "is to transform the way buildings and communities are designed, built and operated, enabling an environmentally friendly, socially responsible, healthy and prosperous environment that improves the quality of life."

Prior to the efforts of organizations like the USGBC, the concept of what constituted a "green building" in the U.S. lacked a credible set of standards. The USGBC's Green Building Rating System has a goal of applying standards and definitions which link the idea of high performance buildings to green construction practices. The program developed by the USGBC is called "Leadership in Energy and Environmental Design" (LEED™). Sustainable technologies are firmly established within the LEED project development process. LEED loosely defines green structures as those that are "healthier, more environmentally responsible and more profitable."[5-10]

MEASUREMENT AND VERIFICATION IS VITAL

Increasingly, measurement and verification (M&V) is being used for green building projects. M&V refers to the process of identifying, measuring and quantifying utility consumption patterns over a period of time. Measurement and verification can be defined as the set of methodologies that are employed to validate and value proposed changes in energy and water consumption patterns that result from an identified intervention; e.g., set of energy conservation measures, over a specified period of time. This process involves the use of monitoring and measurement devices and applies to new construction and existing buildings and facilities.

Measurement and verification methodologies are used for LEED projects, performance based contracts, project commissioning, indoor air quality assessments and for certain project certifications. By establishing the standards and rules for assessment criteria, the concept of measurement and verification is a key component of energy saving performance contracts. In performance contracts, the performance criterion of a project is often linked to guaranteed cost saving that are associated with the facility improvements.

Technologies and methodologies are available to measure, verify and document changes in energy usage. Tools are available in the form of M&V guidelines and protocols that establish standards for primary measurement and verification options, test and measurement approaches, and reporting requirements. Using procedures identified in the guidelines and protocols, a measurement and verification plan is developed to validate savings and to serve as a guide as the process unfolds.

The process of measurement and verification typically involves five primary steps: 1) performing the pre-construction M&V assessment; 2) developing and implementing the M&V Plan; 3) identifying the M&V project baseline; 4) providing a post-implementation report; and 5) providing periodic site inspections and M&V reports.

The theoretical basis for measurement and verification in regard to assessments of resource usage over comparative periods of time can be explained by the following equation:

$$\text{Change in Resource Use}_{(adj)} = \Sigma \text{ Post-Installation Usages}$$

$$+/- \Sigma \text{ Adjustments} - \Sigma \text{ Baseline Usages}$$

Baseline usages represent estimates of "normal" utility usages prior to implementation of any cost saving improvements. Adjustments are changes in resource use that are not impacted by an intervention and are considered exceptional. The term intervention refers to the implementation of a project that disrupts "normal" or "projected" utility usage patterns. Examples of these interventions include electrical demand reduction measures and energy and water conservation measures. Post-installation usage refers to resource consumption after the intervention has been performed. Using this formula, negative changes in resource use represent declines in adjusted usage while positive changes represent increases in adjusted usage.

The International Performance Measurement & Verification Protocol (2007)[5-11] is the most widely used M&V protocol. Its guidelines will be used as an example. The measurement and verification options in the IPMVP provide alternative methodologies to meet the requirements for verifying savings. The most recent version (2007) of the guideline is available on the web site evo-world.org. The four measurement and verification options described in the IPMVP are summarized as follows:

Option A: Partially measured retrofit isolation.

Using Option A, standardized engineering calculations are performed to predict savings using data from manufacturer's factory testing (based on product lab testing by the manufacturer) for a specific energy efficiency measure and a site investigation. Select site measurements are taken to quantify key energy related variables. Variables determined to be uncontrollable can be isolated and stipulated; e.g., stipulating hours of operation for lighting system improvements.

Option B: Retrofit isolation of end use, measured capacity, measured consumption

Option B differs from Option A, as both consumption (usage) and capacity are measured (output). Engineering calculations are performed and retrofit savings are measured by using data from before and after site comparisons; e.g., infrared imaging for a window installation or sub-metering an existing chiller plant.

Option C: Whole meter or main meter approach

Option C involves the use of measurements that are collected by using the main meters. Using available metered utility data or sub-metering, the project building(s) are assessed and compared to baseline energy usage.

Option D: Whole meter or main meter with calibrated simulation

Option D is in many ways similar to Option C. However, an assessment using calibrated simulation (a computer analysis of all relevant variables) of the resultant savings from the installation of the energy measures is performed. Option D is often used for new construction, additions and major renovations.

Depending on site conditions and the technologies being used, each approach has discrete advantages and disadvantages. For example, in cases where facilities have main meters in place, Option C may be preferred. In new construction, Option D is the favored alternative.

With recent advancements in monitoring and measurement technologies, it is possible for energy engineering professionals to log and record most every energy consumption aspect of the conservation measures they implement. Examples include the use of data loggers, infrared thermography, metering equipment, monitors to measure liquid and gaseous flows, heat transfer sensors, air balancing equipment, CO_2 measurement devices and temperature and humidity sensors. Remote monitoring capabilities using direct digital controls (DDC), fiber optic networks and wireless communication technologies are also available.

Measurement and verification costs vary as a function of the methodology, the complexity of the monitoring, the technologies employed, and the period of time that M&V needs to be performed. As applied monitoring technologies evolve and become accepted by the marketplace, costs for installed monitoring equipment will continue to decline as the capabilities of monitoring technologies continue to improve.

GREEN CONSTRUCTION IN SCHOOLS

Schools are important. A "Green School" is a "high-performance" school. More and more, the professionals who design and construct new schools or major school additions are aware of their responsibilities. Between the ages of 5 and 18, students will spend roughly 14,000 hours of their lives in school buildings. According to Eley, high performance schools require green construction practices, provide comfort and a healthy environment for students and staff, and use energy and other resources efficiently and have lower maintenance costs. High performance schools also involve a commissioning process, are environmentally responsive, are safe and secure, and feature stimulating architecture.

To spearhead the "Green Schools" effort, standards are being developed and employed. A number of recently developed manuals are available that provide guidance on how to implement green construction practices. The Sustainable Buildings Industry Council (SBIC) has produced a High Performance Schools Resource and Strategy Guide to show school building owners and operators how they can initiate a process that will result in better buildings—ones that provide students with better learning environments.

For new construction, kindergarten through 12th grade (K-12) school systems in New Jersey, California, and elsewhere have adopted their own sustainable building standards. Kentucky is an example of one state that is developing new standards for "Green and Healthy Schools." The USGBC is developing a LEED rating system for educational structures.[5-12]

New Jersey

New Jersey has codified its construction practices for schools in a document entitled 21st Century Schools Design Manual, developed by the New Jersey Schools Construction Corporation. Its performance objectives are structured to create schools that are healthy and productive, cost effective, educationally effective, sustainable, and community centered. The manual establishes a set of 24 comprehensive design criteria for schools and mandates that design teams consider the following categories of issues in new construction.

Each topic covered in the New Jersey manual provides a set of recommendations and identifies applicable standards. Interestingly, the manual's recommendations are similar to LEED-NC prerequisite requirements and elective credits. Schools meeting the standard can be considered to be high performance, LEED-like facilities—while avoiding the rigor and costs of LEED certification. The manual is a call for integrated design solutions to establish sustainable design as a cost effective means of achieving high performance schools in New Jersey.

California

The Collaborative for High Performance Schools (CHPS) is a nonprofit organization established in 2000 to raise the standards for school facilities in California. Goals include improving the quality of education by facilitating the design of learning environments that are resource efficient, healthy, comfortable and well lit—amenities required for a quality educational experience. The CHPS program for high performance schools has been adopted by 14 school districts in California. The standard was recently updated and reissued as the 2006 CHPS Criteria and applies to new construction, major renovation, and additions to existing school facilities. The CHPS offers a Best Practices Manual that details sustainable practices and resources available for the planning, design, criteria, maintenance and operations, and commissioning of high performance schools.

Kentucky

To implement sustainable technologies to help improve education in schools, Kentucky has developed its "Green and Healthy Schools Program," a voluntary program to encourage green standards for schools.

Twenhofel Middle School in northern Kentucky uses a number of technologies to reduce energy and water consumption. The building shell provides extensive use of insulation and high performance fenestration. The school uses a geothermal heat pump system for heating and cooling. Rainwater collected by a metal roof and drainage system is treated and used for non-potable needs. A central computer control system manages the energy used in the building. In addition, solar panels, installed on the roof, collect energy to generate a portion of the school's electrical requirements.

The building makes extensive use of daylighting in classroom areas. Sensors in the classrooms detect light levels to allow fixtures to adjust light output in response to the daylight available in classrooms. There is a touch screen monitor in the lobby of the school that allows students to monitor information concerning the water collection system, solar output and the geothermal loop. Science programs use the school as a learning laboratory. Students often provide guided tours to explain the sustainable technologies that were incorporated into the design of the facility.

CONCLUSIONS

Buildings are resource intensive in their construction and operation. Buildings are also complex systems. Today, buildings can be constructed with features that allow them to use less energy and consume fewer resources. Developing a green building project is a balancing act and requires a series of tradeoffs. It involves considering how buildings are designed and constructed—at each stage of the project delivery process.

Figure 5-6. Display in Lobby of Twenhofel Middle School, KY

Figure 5-5. Solar Collector Array at Twenhofel Middle School, KY

CSOs will find that standards are constantly evolving. The 2006 International Energy Conservation Code (IECC), for example, requires that certain energy efficient design methodologies be used in construction. The code "addresses the design of energy-efficient building envelopes and installation of energy-efficient mechanical, lighting and power systems through requirements emphasizing performance."[5-13] It is comprehensive and provides regional guidelines with specific requirements for each state in the U.S. New construction materials and products are available that offer new design solutions.

There are many opportunities to include green design features and components in buildings to make them more sustainable. Yet, there are differences in the standards for green construction. While energy assessment systems for buildings (the Energy Star program, for example) typically focus the analysis on source energy, the USGBC's LEED program considers the cost of energy as a primary rating criteria. A number of assessment systems for sustainable buildings are now being used throughout the developed world. LEED-NC is becoming a widely adopted standard for rating newly constructed "green" buildings and projects in the U.S. and elsewhere.

In the U.S., a total of 49 localities and 17 state governments now encourage the use of green building practices, policies and incentives. Their number is growing. The USGBC estimates that 5 percent—almost $10 billion—of current nonresidential construction in the U.S. is seeking certification. According to Richard Fedrizzi of the USGBC, "this movement has created a whole new system of economic development... We are at a tipping point."

Many green building technologies such as high efficiency windows, solar arrays and day-lighting applications are easy to find when walking by or through a building. On the other hand, it is discouraging to owners that many important engineered features of green buildings are hidden from view in the mechanical rooms and spaces not visible to the ordinary visitor. Achievements in sustainable building design often go unnoticed by people who visit, work or study in a green building. Examples of these technologies include computer control systems to manage energy and water use, rainwater collection systems, lighting control systems, under-floor air flow systems, etc. Many green technologies used in structures require a trained eye to observe.

Chapter 6

Auditing For Sustainability

James W. Brown

In 1982 I was asked by the Southwestern Bell Company to help them modify and implement their first corporate wide energy management program. It was called BEMAR, Building Energy Management and Renovation, and it focused almost entirely on surveying facilities and making adjustments to improve efficiency. We called the work that we did *energy auditing* and we are still doing the same thing 28 years later. Although the energy efficiency industry has matured greatly over the years, the energy audit re-

BACK IN '82 WE CALLED IT "ENERGY AUDITING"

mains as the backbone of all additional components and aspects of our field today.

I may stand alone in this opinion, but I believe that the tremendous success of the energy auditing business has paved the way for the sustainable design practice that has emerged over the last decade. Those original energy audits discovered some significant renovation opportunities, most being what we called low hanging fruit, that not only saved energy but they saved money as well.

Upon the building block of these energy audits came the concept of "investment grade auditing." The IGA moved energy auditing from the world of general concepts supported by rule-of-thumb efficiency data and estimated cost projections into a higher level of engineering analysis requiring detailed calculation processes to support their thoroughly documented conclusions. Without evolving into the higher plane of the investment grade audit (IGA), most significant energy efficiency projects would not have been installed and the energy efficiency industry would have slowly died away.

Then came "Master Planning for Energy Efficiency." These master plans began by auditing facilities, using the more believable results obtained through the IGA, to see what was happening in the building and what it would cost to fix it, then planning for long-term upgrades and efficiency improvements that extended beyond HVAC and electrical modifications, integrating such items as building envelope components and facility orientation.

Now, we have sustainability departments. What is done within those departments has been described in other chapters within this book, but in my opinion, one of the foremost goals of the chief sustainability officer (CSO) must be to integrate the past achievements of our energy industry into the world of sustainability.

At the end of the day, it was the cost savings achieved by those first energy audits that provided funds for facility improvements that could not be justified on a purely financial basis. As a result, the industry remained on the scene…and it grew. Now we find ourselves in a world of construction where sustainability is the mantra of every consultant and contractor on the job. The funny thing is that sustainability means that the building is supposed to stay around for a long time, and if it's going to be here a long time, we need to keep it operating at peak levels. How do you do that? Through audits! So, even as our industry grows and matures, we find that the backbone has not changed. We sustain our buildings by continually auditing our buildings. So before I get into this chapter on auditing for sustainability, let me pass on an old Icelandic proverb to building owners and operators, "He who lives without discipline, dies without honor." That proverb is as true for your buildings as it is for your lives. Auditing requires discipline. It takes time, commitment, money and discipline. But if not applied, all the money you are putting into a sustainable building will be for naught.

The entire premise behind the concept of auditing for sustainability revolves around the central idea that a sustainable building has been handed over to the owner and that it has been constructed (if new) or renovated (if not so new) so that it can be sustained. When this criteria is met, *then* we have a facility that can be routinely analyzed, surveyed, tweaked and adjusted to continue the original and/or renovated sustainable status.

As we look around at our sustainability needs today, it is increasingly obvious that the energy audit methodology established the protocol for the water management audit, the sustainability audit, etc. Those, who can deliver a quality investment grade energy audit, can broaden their approach to observe, analyze, calculate and recommend other measures which will enhance the quality of a facility and its work environment.

When we make the declaration that we are designing and constructing sustainable buildings, those who hear our declaration are undoubtedly assuming that we are building facilities that we want to have around fifty to a hundred years from now; buildings that will offer productive and effective service year after year for all the future generations that will eventually occupy them. In effect, we are telling those who will listen that we are creating facilities that can be maintained and renovated over the years to provide service to the occupants fifty years from now equaling the service it provides on the day of its grand opening. When asked what describes the difference between our old designs and those we now proclaim to be our standard of design we simply state that we now build buildings that are "sustainable."

The *ASHRAE Handbook of Fundamentals* defines "sustainability" as "providing for the needs of the present without detracting from the ability to fulfill the needs of the future." In fact, sustainability is really all about the future. Obviously, the desires for the present must be met. The current reason for the building and the desires of those paying for the building must be inserted as variables in the equation. But when the design is labeled sustainable, present values run a distant second to a well preserved, high quality future.

As a result, those variables considered all important during construction projects of the past should be given less consideration in the evaluation of what, and what not to do in today's design/construct process. For a long time now, it has been considered to be good business to evaluate the long-term value of the current investment using business calculations, like life-cycle costing. The end result of such analyses, however, tends to weaken the importance of the initial cost, propelling issues such as maintenance expense to the forefront of the decision making process. In fact, when using any long term evaluation tool, the further you look into the future, the less important the initial construction cost becomes. Conclusion: *the result of any design claiming to be sustainable must, by definition, favor long term values over short term cost.*

As I ponder this conclusion, a heavy fog settles over my engineer slanted brain because lying here in front of me is our design team's latest new construction drawings. It's very presence on my desk forces me to face the reality of the real world that I live in. When I adjust my reading glasses to truly see the results of our latest effort, I have to admit that most of the pre-design "green" talk got lost somewhere in the inevitable value-engineering discussions. Some of the non-efficiencies I see are not even money related items. They simply represent systems that the owner understands, and because he understands the old way, he feels more confident that they will not present problems in the future. Try as we might, we simply couldn't convince the decision makers that water cooled chillers, condensing boilers and displacement ventilation air distribution systems are more efficient than the old designs they have become accustomed to.

First cost has once again triumphed. Truth be told, no matter how many times we say the word "sustainable," first cost still reigns as the supreme variable in the construction equation. As a result, we have begun to use that truism in our arsenal for convincing clients to institute an effective auditing program. I won't bore you with the sales pitch, but the bottom line is that auditing the facility is like putting oil in the car... if you want it to keep running smoothly, you need to glance at the dipstick now and then.

IT'S LIKE PUTTING OIL IN THE CAR

Most older buildings can be renovated to meet the owner's needs *and* can remain a viable location for their occupants *if* they will renovate it with sustainability in mind *and keep a close eye on it throughout its useful lifespan.* That last part requires consistent auditing of the facility. If the owner will buy into that concept, then both new and existing facility construction projects can honestly bear the sustainable label.

Before getting into this concept of sustainable auditing procedures, I feel almost obligated to say that experience has led me to the conclusion that not all buildings *can* be brought up to a level that even the most gracious auditor would call a sustainable facility. They are old, inefficient and typically uncomfortable places to work. They leak outside air from every joint and crevice, are cooled by chillers with CFC-non-grata refrigerants, and have so much asbestos and lead paint that it would cost far more to make it sustainable than it would to tear it down, and even tearing it down would cost a fortune.

Fortunately, those buildings have become the minority in the USA today. A large majority of buildings now being analyzed by our firm's energy consulting department can be brought into the eco-friendly family and are benefitting from the huge bundles of stimulus money being circulated around the country these days. Based upon the direction that the stimulus funds are going, our country seems to agree that saving an older building is about the best thing we can do for our environment. Just think of all the brick, steel, concrete and plastic that isn't being dumped in our landfills these days simply because we have determined to upgrade those buildings that can be brought up to desired standards rather than focusing all of our construction funding on tearing down the old and raising up the new.

Our office is headquartered in a building constructed in 1876 and is located on the Main Street of our growing community. On the wall of the building is a plaque stating that it was deemed to be an historic facility by the State Historical Survey Committee in 1970. Included on the plaque are the words "The Old Broom Factory" depicting its original function and reminding the community that this is the site where the broom that won first prize at the St. Louis World's Fair in 1904 was manufactured. Our local citizens have given the building a place of honor in the city register. With that honor comes all rights, privileges and tax deductions typically awarded to community icons. In return for the honor and the tax deductions, the community requires (yes, *requires*) us to "sustain" it, which by definition means "to keep it up, keep it going," in the manner of its original condition and usefulness. It isn't about the brooms, or the general store, or the school that once inhabited the building...it's about the building...keeping it functional, usable and productive. Somehow, sustaining this old building brings pride to the community and makes us feel that people can leave an imprint on the earth that lasts long after they leave this life.

Now, let's talk about today's buildings. There is little doubt that functional facilities are springing up all across this land, facilities that are designed for a specific function and should serve that function well for as long as the need remains. The operative word here is *should*.

Technological advances change the shape and operating requirements of the equipment housed within the building, equipment that is needed to produce the company's end products. These rapid advances in technology often make the building inadequate for the current product being produced by the building occupant. In addition, building materials believed to be safe during the time of the construction are now so ardently regulated and so expensive to remove that it is frequently cheaper to move out and build a new place to work. In short, a lot of the construction going on today revolves around function and convenience and today's suitability of that building for its current purposes. Thus, true life experience shows us that sustainability isn't even the second most important building consideration today. The first is "first cost" and the second is "building function."

The result of this admittedly limited survey is that we have not yet entered into that oft spoken and profusely praised era of construction that is primarily focused on the longevity or sustainability of the building.

As a result, surprising as it should be to hear it said, especially in a world with magazine and university classes alike filled to the rafters with *sustainable* rhetoric, the simple fact is that "they *still* don't build 'em like they used to."

Being one of the designers of today's buildings, I hasten to add that the end result that is still dominant in today's construction is rarely what we set out to accomplish. As a mechanical, electrical and plumbing (MEP) design firm, specializing in energy efficiency, we are extremely proud of our designs and (for the most part), the buildings we have helped construct. But I don't believe that anyone on the design/construct teams we have been involved with would say that we are producing facilities that will adequately serve the need fifty years from now, much less the 100-year life so commonly offered by previous generations.

The defining line then, when discussing the "to be or not to be" of sustainable buildings, lies within the attitude of construction and the owner's commitment to adequate maintenance, not only at the beginning but throughout the life of that facility.

You are thinking that this is a strange way to more forward with a chapter on building auditing, aren't you? Audits are, after all, primary tools of sustainability. So if buildings are constructed with short term function as the overriding goal, then keeping the building at peak operating condition isn't as important as it would be if the lifespan of the building were the prevailing goal. Well, the truth lies exactly in the opposite corner. Cost-cutting measures implemented at the outset create an even greater urgency for up-keep. In fact, almost every dollar saved during the construction process results in four more dollars needed in the maintenance budget over the life of the building. How do I know this? Because almost every study I read about the life cycle cost of a building states that 80 percent of that cost will be spent on maintenance and operations. Thus, it is safe to assume that cheaper buildings will require fatter maintenance budgets.

In another chapter we will discover that owners who avoid the "first cost" of commissioning will almost certainly operate for years, possibly throughout the life of the facility, with improperly functioning, or should I say non-functioning systems. It has been documented by study after study that retro-commissioning; i.e., commissioning a building that has never before been commissioned, offers the best opportunity for savings in that building. Think about that for a moment. Hopefully, you won't have to think long before coming to the conclusion that the longer your building goes un-commissioned, the more wasted money you are sending to the utility companies...more about that later.

Because you are reading this book, it is presumed that you want to sustain your building. That being the case, there is actually only one really appropriate path for you to choose: set a course that obtains and maintains the highest possible efficiency with the lowest possible environmental impact.

When Wal-Mart was proclaimed "environmentalist of the year" in a recent "green" newsletter, the basis of their selection was justified by their movement toward energy efficiency concepts, indoor air quality improvements, and selected water management decisions. When commercial businesses, such as Wal-Mart, begin considering the impact that they have had on our national environmental picture, and set their corporate minds toward wider and broader environmentally friendly construction and operation, we can conclude two things without hesitation: *first,* is that America's big business has made it their business to protect our environment and reduce our country's enormous consumption of energy; *second,* we can conclude that sustainability is good business. We have before us the nation's leading commercial giant with its sights set on "going sustainable." Note that I didn't say that they were "going green," because the *green* word suggests environmentally friendly at any cost, while the *sustainable* word integrates good business into the equation. Though I have complete respect for Wal-Mart, I seriously doubt that they made their environmentally friendly decision without looking into the value added to their construction and operations budgets. The bottom line is that when "green" becomes sustainable; "green" becomes profitable.

Even adding profit to environmental improvement, however, doesn't move people into a green lifestyle. In the February 13-14,

2010 edition of the *Wall Street Journal* there was an article entitled "Even Boulder Finds It Isn't Easy Going Green." In the article, we find that at some point the city officials of Boulder, Colorado, set out to become a role model in the fight against global warming. But what they found was that governmental good intentions don't always alter the personal practices of the citizens. In fact, one of the senior research fellows at the University of Colorado in Boulder was quoted to say, "What we've found is that for the *vast majority of people*, it's exceedingly difficult to get them to do much of anything." (*italics* added for emphasis)

I added the emphasis to the above quote in order to reassert a well known fact of life: convenience trumps conviction.

I wish I could say that this little adage had been discredited by our firm's auditing department. I wish I could say that the behavioral modification recommendations included in every one of our audit reports had proven to be the easiest of all recommendations to implement. But if I said it, I would be lying, and you would immediately close the book and toss it in the trash. The real truth is that occupant behavior, which by the way represents the single largest energy savings potential in our country is, as the City of Boulder will attest, "exceedingly difficult" to change. So as we proceed into deeper discussion about auditing, let me issue the warning that the greatest risk we face as we proclaim the cost savings available in facilities is that the people will (not may…will) find ways to modify our estimated results.

Dr. Hansen and I addressed this issue in far greater detail in the book we authored six years ago, the *Investment Grade Energy Audit* (The Fairmont Press, 2004) and as far as I can tell, it hasn't changed. Our firm generates over 200 audits each year, most involving public entities, and we find over and over again that the risk to the veracity of our reports doesn't lie in the technical calculations or the detailed construction costs that we so thoroughly evaluate. The risk lies within the way our recommended measures are perceived by, and adopted by, the occupants of the building. In a recent survey of public school energy managers taken by a Texas based engineering firm, it was discovered that the performance of the equipment purchased ranked highest in their perception of value, and LEED (or any other sustainable certification) ranked tenth...out of ten suggested categories!

As a result, the most effective task, as we attempt to sustain the sustainable building for the CSO and his/her consultants, will be to evaluate the occupant, communicate the need behind the revision, and participate in the continuing implementation of the renovation. In retrospect, we should have implemented this strategy long before the idea of sustainable auditing was ever conceived. We didn't, but perhaps we can begin now. It is the only way we will ever be able to claim success within the sustainable construction industry.

We have entitled this chapter "Auditing for Sustainability" for one simple reason. Without consistent, effective audits of all primary building components, all the variables that make up the definition of sustainability will eventually erode.

In the "Energy Use and Management" chapter of ASHRAE's *2007 HVAC Applications Handbook* is a ten-step grid for introducing the energy management process into your organization. Six of those steps basically describe an energy auditing program.

Let me lay them out for you:

- Designate and define energy manager responsibilities and qualifications
- Establish communications
- Establish energy accounting system
- Analyze energy data
- Perform energy surveys and audits
- Improve discretionary operations
- Evaluate energy conservation opportunities and prioritize
- Implement energy conservation measures
- Monitor results
- Evaluate success and establish new goals

As the energy auditing process has evolved over the years, the tasks provided by any comprehensive report include each of the above listed tasks from analyzing energy data through monitoring of results. In fact, our firm has been called upon to assist the client implement the entire ten step program while operating under a contract originally intended to simply provide a few energy audits.

I won't go into a description of each of the components. The *ASHRAE Handbook* does a far more thorough job than I could do within these pages. The bottom line is that the energy efficiency industry has finally brought enough value into the process of building operations that we have earned a "place at the table" (as my co-author Dr. Hansen is fond of saying). We have become an important component within the owner's definition of a successful project. As such, it is not, or certainly should not be, within our business strategy to simply hand clients our efficiency reports and quietly walk away. The application of the report's findings is ours to implement. If we haven't brought our services up to the point of implementation *and* verification of our proclaimed savings; if analyzing and surveying are typically all that we do; then we haven't done our part in bringing the energy consulting industry up to a believable profession. Listen now as I beat my drum: *energy consultants today need to learn how to assert the value of their work.* They need to be as insistent about the value of their findings as the structural engineer is about the beam sizes selected to support the building. We need to be confident that our recommendations will serve the client's needs and stop being content to produce a report that gathers dust on his shelf.

Let me give you an example: Our firm has public school clients that actually plan their future bond program projects around the master plans that we produced for them. Those plans were based upon thorough facility audits with energy efficiency being the centerpiece. In fact, the cost savings projections entered into those master plans are used to support the financial viability of the projects we recommended, as well as being counted on to help the school install needed projects with no associated payback. I don't know how to say this humbly, so let me just say it out loud: Gentlemen, when you get to this point, you know you're in the game!

So we have started with the ending. Bringing auditing to the point achieved with these clients is as far as I can see. Maybe there are higher goals and loftier achievements to be accomplished, but I don't know what they are. At this point all I can give you is what we believe to be the definition of success, then point to the process for getting there.

Obviously, not all clients start you out at the highest level of audits. There is usually that testing period for you to show them that you know your stuff. So, once again, ASHRAE shows us the steps toward the ultimate auditing contracts – whether they are energy audits, water management audits, or the more sophisticated sustainability audits.

Again in ASHRAE's Energy Use and Management chapter, we find three levels of energy audits.

Level 1 is called the "*Walk Through Assessment.*" Although this level is typically delegated toward finding lower cost projects, generally of the O&M variety, it is also a very useful confidence building procedure, allowing the auditor to assess the client's facilities and offer suggestions as to which ones provide the greatest opportunities for savings. The audit does not have to be limited to low cost/no cost projects. In fact, it makes a lot of sense to add a laundry list of potential capital expense savings opportunities within these reports. The twist that brings the confidence for the client, however, is to select a few of the projects with the most savings potential and provide a greater degree of detail both for energy cost savings and project implementation costs. This is the part that generally propels the audit up to the next level.

The Scoping Audit, a Level 1 audit in the performance contracting industry, is also used to determine if the potential energy and water savings opportunities justify more intensive (and costly) work.

In today's audit climate, there is yet another opportunity for use of these Level 1 audits. They can be used to determine if the sustainable aspects of the building's design are still functioning as designed. Any building that has been certified as green or sustainable by an accredited agency will have a list of the items inserted into the design that garnered points toward the final certification award. As a result, the Level 1 audit can be useful as a tool to insure that the sustainable credentials are still merited, and if not, what needs to be done to bring it back to its original status.

It should be pointed out that many of the agencies offering sustainable credentials have created programs that tend to offer more points for efficient HVAC, controls and electrical service than for any other building component. That fits right into the services included in these Level 1 audits. For example, the Collaborative for High Performance Schools (CHPS) as modified for application in the State of Texas, has 129 points available from the various construction categories with 67 of those points offered for Water Conservation (non-irrigation), Energy Efficiency, and Indoor Environmental Quality. The MEP installations produce all of these 67 points. As a result, there is great potential for energy auditors to serve as a verifying group and inspect for continuing operation of the items included within the TX-CHPS score card. On the other hand, a building could be certified as sustainable without any real emphasis on energy efficiency. All that is absolutely required is that the building meets minimum efficiency performance standards (a point that almost all new building designs have achieved these days). In fact, the higher level offered by TX-CHPS (Verified Level) requires only 2 points be gained from the Energy Efficiency category. If you keep this in mind, as a CSO you should expect and get an even more substantive product from your consultant(s).

So it may be a good idea to alter your standard energy auditing format to go beyond energy consuming equipment renovations. Maybe it would be a good idea to include within your survey tool chest, inspections for indoor air quality, water consumption retrofits, hazardous waste analyses, and recycling opportunities. These are all portions of a sustainable facility and all need to be inspected on a routine basis.

When surveying the building's energy consumers, look around to see if the EnergyStar appliances installed during the construction process are still in operation. Those commercial dishwashers save 25 percent in energy and 25 percent of the water commonly used by this type unit; commercial hot food holding cabinets are 60 percent more efficient than their normal competitors; even copiers and fax machines provide 25 percent energy savings over non-EnergyStar units. Today's sustainable auditor makes it a point to see if those units are still there, or if they have been replaced with less efficient equipment.

Level 2 is called the *"Energy Survey and Analysis"* level. It differs from Level 1 in that it offers more intensive analysis and more detailed cost savings and construction cost information. Most projects with total implementation costs of less than $100,000 can be initiated confidently after this Level 2 audit is completed. As a precaution, we find it advisable to insert a review process into this level of audit because it isn't necessarily an "investment grade" audit and may need the opinion of another experienced auditor before giving the client full assurance of its credibility.

Level 3 is called *"Detailed Analysis of Capital-Intensive Modifications"* (a.k.a. investment grade energy audits.) Comparing this level to Level 2, the primary difference lies in the degree of confidence in the variables included within the analysis. This is the level that requires field measurements, energy modeling, lighting pattern and distribution analysis, water distribution modeling, and any other degree of detail that is needed to minimize the opportunity for calculation error. This is the level that the industry should embrace as "investment grade." In fact, it is the level required for completion by MEP engineers when applying for the credits leading toward loftier sustainable certifications.

AUDIT QUALITY

Before concluding this chapter, I want to return to that basic concept previously discussed which is the success-formula for any successful auditing venture: *"Don't forget the occupants."* No matter which level of audit you are performing, the end results will not lie in your experience, intelligence or attention to detail. The utility bill will be determined by those who inhabit the building. This has been said far too many times over the years, and has been disregarded far too many times in the process. The successful audit includes a step not mentioned in technical guide books. That step involves communicating with the building occupants. It is worth noting again that in our last collaborative effort, the *Investment Grade Energy Audit*, Dr. Hansen and I had a whole chapter, "Weighing Human Behavior" devoted to this topic. It is still worth a read today. Considering the "people factor" is particularly important when you discover that something included on the sustainable checklist that gained the building the distinction of being an environmental asset has been discontinued due to occupant disapproval.

One last reference to the *ASHRAE Handbook*. The sixth step in the stated energy management procedure is to "improve discretionary operations." As we have matured in our specialty field of energy efficiency, we have concluded that the most important element within the term "discretionary" is the discretion of the building occupants. Our conviction at this point in our history is to once again turn our efforts toward occupant participation in the cause for energy efficiency.

For the record, there are a few energy efficiency firms in the country that base their entire efficiency effort toward the adjustment of the behavior of the occupants. Most of those firms go so far as to guarantee the savings that will result from those behavioral adjustments. Ours is not one of those firms. Why? Because although we are convinced that significant savings exist in this arena, we are equally convinced that one disgruntled school principal or one uncomfortable primary tenant can torpedo those results almost overnight. Further, without a specified level of support and training/re-training, the behavior (and the savings) erode over time.

It is at that point that we disagree with them whole heartedly! Our industry has come a long way toward gaining credibility in the technical side of energy efficiency. However, although we can control equipment operating parameters, lights and equipment on/off times, we cannot control occupant attitudes or actions. I recently saw a proposal from one of these firms offering to save the client 30 percent of its utility bill through "behavioral modifications" of the occupants. This achievement was to be provided for a fee of only $300,000! I know the client and I know the excellence of their energy management department. The promised savings would not have been achieved. And charging that client for savings that were not accomplished because the firm wasn't able to deliver its promised behavioral adjustments is disgraceful, not to mention being extremely detrimental to our energy consulting industry.

Even so, the *true* professionals in our industry must re-focus on the impact of the occupant on our desired results. We learned it a long time ago, and now it's time to learn it again: *If a building is ever going to sustain its sustainability credentials, it will have to have the support of those with the power to kill-the-deal.*

One final note: In January of 2010, ASHRAE launched Standard 189.1 *"Standard for the Design of High Performance, Green Buildings Except Low-Rise Residential Buildings."* This is a standard that defines minimum performance requirements for green buildings. It isn't a design guide, nor is it another green rating system. It was simply published to complement current rating and certification programs. It's complementary suggestions support improved design and operations by addressing facility sites, water efficiency, energy efficiency, IAQ, atmosphere/materials/resources, and construction and operations plans. It's predicted impact will generate around 30 percent energy savings when compared to Standard 90.1-2007, and 40 percent water savings compared to EPAct 1992. This standard should be studied by every CSO, sustainability auditing organization, and all those who want to see their recommendations become implemented.

The industry keeps changing… and so should we.

Chapter 7

Commissioning:
At The Heart
Of Sustainability

James W. Brown

On a recent trip, my wife took me to an opera featuring a soloist, a mezzo-soprano, who sang her way through twenty-five songs (yes, I counted them) that somehow represented the story of her life and career. As an engineer at the theater, I did what all sound-minded engineers would do during such sophisticated and life changing moments—I began inspecting the building, at least the part that I could see without turning my head too far in any particular direction (remember, I said my wife was with me). I inspected the unusual acoustical panels suspended from the ceiling, the lighting system (trying diligently to determine which of those lights were the ones called "tormenters"), and the sound system. Needless to say, I felt more than a little out of place, a fact I accentuate by saying that it was song number eighteen before I heard the first understandable words uttered from the stage.

Now, if you will allow me to bring that little piece of experience into the world where chief sustainability officers (CSOs) live, please heed this advice: "Sing your arias in the language of the audience."

I can't tell you how many meetings I have attended where the sustainability guru waited until the "eighteenth song" to say anything that touched upon the responsibility of the design team. There are no engineers that I know who wouldn't appreciate it if someone would save the ozone layer and slow the pace of glacier melting, but when meeting with the building owner and the construction team, focus your comments on the sustainable concepts that the team can address! CSOs need to know beforehand which sustainability certification they plan to pursue and focus their comments on the items that will produce the building they want and the points that result in certification.

Prior to this chapter, you read quite a bit about the impact that a sustainability officer can have on global conditions that are not just impacting, but actually threatening, the world we live in. Hopefully, every CSO in every corporation is stirred to action to do whatever it takes to stem the tide of man-made activities having such potentially disastrous effects upon our planet. The designers of the world salute you. However, when you come to a construction meeting, come prepared to sing our song…tell us what we need to do…tell us what the building owner desires in the way of sustainability. Discuss LEED criteria, help us evaluate which points we should pursue, tell us about selected EnergyStar purchases and federal funds available to us if we achieve the desired level of "green." In short, tell us what our "sustained" building will look like when we finish the construction. Then, tell us what is going to be done to ensure that all the sustainable concepts desired will be installed properly and work effectively when we turn the keys over to the building owner.

You know where I'm going here. There is only one way to offer that assurance. The building must be commissioned!

Let me join with everyone who has ever authored a book on commissioning and provide what I believe to be a workable definition of the task. My preferred definition was written by Brooks Energy and Sustainability Laboratory in a study funded by the Texas State Energy Conservation Office in May 2007. The resulting report was entitled "The Viability of Commissioning in New School Construction" and it contained these words:

> *"Commissioning is a well known process whereby commissioning professionals verify building systems/components as well as functionality and operation against owners and designers intent. The American Society of*

Heating, Refrigerating and Air-Conditioning Engineers (ASHRAE) de-
fines commissioning as "a quality oriented process for achieving, verifying,
and documenting that the performance of facilities, systems, and assem-
blies meets defined objectives and criteria."

The report compares two test-case, new-construction projects;
one of the buildings was commissioned during the design and construction process, and the other was not. To summarize the results
of this analysis, let me report some of the study's findings:

During the construction process, there was only one-third
the number of change orders issued in the commissioned building as in the non-commissioned building, and the cost of those
change orders was 45 percent less in the commissioned building.
No specific cost data were provided, but experience tells me that
this result alone would more than have paid the fee charged by
the commissioning agent.

After the buildings were completed, the commissioned
building had 50 percent fewer mechanical, electrical or plumbing
(MEP) related work orders during the first six months of operation, and the cost of materials and labor for the resulting repairs
was 60 percent less than in the non-commissioned facility. Once
again, my hunch is that these savings were greater than the commissioning fee would have been.

Finally, the cost of energy within the commissioned building was $0.016/sf less than for the non-commissioned building.
That's using *today's* cost of energy. If someone would check out
the total savings over the next ten years including rate increases,
occupancy increases and building additions, I wonder how much
this commissioning induced improvement will reduce the owner's operating budget.

Is commissioning an essential part of today's construction
process? It is if you want a cost effective, sustainable facility.

Let me give you a simplified list of issues discovered by our
firm's commissioning department during the past three years
and allow you to decide.

PROGRAMMING AND SCHEMATIC DESIGN PHASE

Little emphasis for short-term and long-term energy savings by the architect/engineer (A/E) regarding heating, ventilation & air conditioning (HVAC) & lighting systems (either disregarding the points and efficiency gained through potential energy credits and designing/installing rooftop units (RTUs) and higher wattage lamps for faster design turnaround times, or passively allowing the value engineering process to revert to lowest first cost systems).

Design Phase
Some of the typical items found in the design include:

- Ductwork undersized or poor layout for duct routing

- Poor or no controls and/or stated sequences of operation for HVAC & lighting systems

- Missing expansion tanks and water treatment systems

- Electrical power for HVAC & lighting equipment missing from the plans

- Missing drains and drain/vent lines for water side equipment and fixtures

- Life safety systems not addressed (at all) during design (FA/fire sprinkler/smoke control)

- Missing specification sections for MEP systems shown on the drawings

- Insufficient means of access to and around HVAC and electrical equipment (servicing or removal and replacement of these systems would require the building walls to be removed or modified)

- No climate control for occupied spaces (missing HVAC for spaces that need heating/cooling)

- Lack of coordination between the design team for MEP system installations (sewer lines shown to be routed through grade beams; ducts shown to be routed through windows and steel structural components; grills and FA devices shown to be installed behind furniture).

After the design is complete and ground breaking ceremonies have ended, the commissioning agent *must be* prepared to spend a significant portion of his/her day on the construction site. This is where the rubber meets the road for the entire commissioning process. It is the reason for many of the commissioning agent's pre-construction hours inserting commissioning language and requirements into the specifications.

If the commissioning agent does his job correctly, there will probably be some point during the construction phase where the owner of the building is overheard reciting prayers to the creator of the earth, and all its residual building materials, for the good fortune of having hired this commissioning agent!

Let me continue the firm's findings with a list of rather significant issues discovered during the building's construction phase.

Construction Phase

- Damaged underground utilities were being covered up without proper repair

- Chilled water/hot water (CHW/HW) piping testing not performed per specifications and found to have leaks during commissioning agent-requested pressure tests prior to pouring the pad and covering them up

- Underground piping flanges not insulated or properly torqued before cover up or pad pour

- Missing J-boxes, conduits and block outs for building FA/telecom/security/MEP items scheduled to be installed in tilt-wall

- Missing conduits for electrical/telecom/future systems during field inspection prior to cover up

- Building steel not secured to the structural support wall for second level floor space

- Cracks in tilt wall requiring structural support plates to be installed to prevent building shifting

- Temporary caps left on underground utility lines (water & electrical)

- Incorrect insulation installed during dry-in of building, which would have resulted in reduced energy savings to the owner and increased load on the HVAC system

- Faulty or missing control relays and control devices for HVAC and lighting systems

- Faulty programming of the direct digital control (DDC) control logic

- Control valves, dampers, sensors installed backwards or on the wrong air/water lines or missing altogether

- Missing light fixtures and emergency lights not operational

- No override function during unoccupied hours

- Missing emergency power off (EPO) switches for disconnect of mechanical/electrical systems for emergency shutdown

- Missing motor starters, disconnects and variable frequency drives (VFDs) for HVAC systems

- Equipment service clearance not maintained preventing access and servicing of equipment

- Piping installed above ceiling found crimped/crushed, which would result in loss of flow to HVAC systems, test/adjust/balance (TAB) issues and occupant comfort issues

- Below-grade piping not installed for plumbing fixtures and floor drains during rough-in

- Missing trap primers during rough in of plumbing lines below grade

- Missing by-pass for CHW loop

- Cooling tower support beams found to be at variance with height requirements per construction documents

- Cooling tower condenser water loop bypass valve installed backwards

- Incorrect control settings on chillers and boilers

- No control of building systems during construction (possibly resulting in unnecessary peak demand charges to the owner and reduction of life expectancy after building opening).

Some of these items in and of themselves may not qualify as "significant" until the final impact is considered, such as finished slabs being cut up, beams being modified, duct work being removed and replaced, roof drains rerouted, building walls modified, pavement erosion, occupant injury, unnecessary shutdowns of building systems, life safety systems failure, continuous comfort issues and reductions in expected energy savings to the owner.

There are a thousand other issues, but these are some of the issues that if left unchecked, will certainly have a negative impact on the owner's staff, operating budget and facility occupants.

Re- or Retro-commissioning—
Make It Work Efficiently, Effectively and Productively

After the sustainable building is constructed and has been operating for a year or two, another equally valuable commissioning process needs to be implemented, a program designed to keep the sustainable building operating as planned; i.e., re-commissioning for previously commissioned buildings or retro-commissioned if never before commissioned. A truly sustainable building means commissioning is never over.

IT AIN'T OVER

Table 7-1 offers some typical results from work provided for our independent school districts (ISD) clients.

The items we discovered during the site surveys that made these retro-commissioning savings possible are typified below:

- Check the energy management control system (EMCS) programming—line by line—make sure it runs when it should and doesn't run when it shouldn't.
  - — Compare what they think is happening to what really happens.
  - — Set up the trend log program and look at how the equipment really operates.
- Check the sequence of operation programmed into the schedules. Don't settle for generalized programming.
  - — Make sure compressors and fans, dampers and valves operate at the most effective time for the area served or for the service provided.
- Set the HVAC system ON time to match the ability of the system, not the experience of the operator.
- Set the OFF time to match the facility U-value and coast toward the end of the day using fans instead of compressors.

Table 7-1. Sample Commissioning ROI

Retro-Commissioning Project		Energy Cost Savings	Installed Cost	Payback Period
"A" ISD				
1. HS		$101,019	$45,100	0.4
"B" ISD				
1. ES	2. ES	$36,605	$34,997	1.0
3. ES	4. ES			
1. MS	2. MS	$28,657	$45,735	1.6
"C" ISD				
1. ES	2. ES	$40,768	$30,876	0.8
3. ES	4. ES			
1. MS	2. MS	$51,399	$27,437	0.5
"D" ISD				
1. ES	2. ES	$105,835	$24,100	0.2
3. ES	4. ES			
5. ES				
1. HS		$21,575	$15,860	0.7
"E" ISD				
1. HS		$13,049	$3,870	0.3
2. Academy		$47,953	$38,100	0.8
TOTAL		**$446,860**	**$266,075**	**Avg. 0.6**

- Review work orders to find the equipment with the highest maintenance costs, and add this to the renovation project.
- Determine if preventive maintenance is being done—preventive maintenance saves money!
- Provide training needed to sustain efficiency of the systems analyzed.
- Install a facility maintenance program in order to track the cost of repair for primary energy consuming systems.
- Analyze utility rates. If there are better rates available, contact the utility provider and discuss the procedure required to shift to that lower rate.
- Review and recalculate monthly bills for errors, then do it again using alternate rate schedules and see if changing rates would reduce cost.
  — Adjust operations to match rate.
  — If your area has deregulated utility suppliers, get information from their representatives and be ready to renegotiate.

In Chapter 1, Dr. Hansen said that the chief sustainability officer needs to "worry globally and act locally." That is exactly what has occurred as organizations such as the Collaborative for High Performance Schools (CHPS) work with local design and construction specialists to modify the manuals of sustainability to fit the local environment.

In 2009, several Texas-based professionals were asked to participate in committees aimed at molding the generally established CHPS practices to fit our state. The result was the TX-CHPS manual, which has already begun to have substantial impact on the practical sustainability of new and renovated schools across the state.

As a member of the technical committee for this manual, it soon became apparent to me that the point system developed for certification of our schools was as greatly impacted by the experiences and desires of the team members as it was upon the true benefit of the recommended actions. Some members saw detailed and continuing commissioning as a vital ingredient of sustainability while others believed that a high quality testing and balancing procedure would produce the same end results. Our experience in the commissioning world has led to a strong support of the former viewpoint. Let me give one conclusive example.

While surveying a large school campus about six months after it had opened, we discovered that a majority of the rotating thermal wheels in the dedicated outside air system (DOAS) compartments were not providing adequate humidity control. The problem: the belt connecting the wheel to the motor had "fallen" to the floor. The amazing thing about this discovery is that most of the belts had fallen right back into the cardboard belt wrapper that they were shipped in. The test and balance (T&B) report showed that the unit airflow was correct but did not measure or address the incoming humidity levels. It seems that sustainability efforts can be relatively easily "unplugged."

It is precisely this kind of discovery that has led school districts like Houston ISD to make it a policy that all new construction projects will pursue LEED certification. It takes guts to make such an announcement, especially in these times when school finances are in such a mess, but the HISD board of trustees saw the value of sustainability and has also seen the complications prevalent in buildings that are not commissioned during construction.

During our auditing experiences, we have found such a wide variety of items that could have been discovered and corrected during the construction process that we simply can't imagine future construction programs with no commissioning involvement. Things like damper linkages missing on outside air dampers, valves piped backwards, meters wired backwards, and empty (yes, completely empty) VFD cabinets. Things like these have such a major impact on the building's environment; yet, they were not installed correctly and never worked.

The current situation we face goes something like this:

• Is sustainable construction a politically correct decision? *Yes*

• Does it always improve efficiency of the building? *One would hope so, but it's not always true*

- Is it an expensive decision? *There are added fees associated with the practice, but when evaluating the overall construction expense at the end of the project, those fees are often (if not generally) offset by the cost savings that can be generated by the commissioning program.*

So, let's look a bit more intently into the way commissioning is supposed to work in your projects.

First and foremost, enthusiastic support by the building owner is needed to achieve the commissioning goals and accomplish desired activities. The CSO, the design team and the contractors need to be instructed that commissioning of the facility, and/or specific systems within that facility, is a desired component of the construction process and that the commissioning agent is being "commissioned" to represent the owner's interests. They also need to be made aware of the fact that the commissioning agent's comments and/or advice will be taken seriously.

If the project is, or contains parts that are ultra-critical, such that project failure is unacceptable, commissioning as an overlay for added insurance may be a great investment. In other words, if the risk of failure is intolerable, then the cost to mitigate that risk is usually warranted.

If the owner considers this project to be critical to the mission of his operation, commissioning will be an excellent tool to help ensure success. Note that a commissioning scope can always be confined to certain areas where the concern for risk management is greatest; e.g., a computer data center in a specialty office building. It stands to reason that another vital directive needed from the CSO is a determination of where *quality* fits in the projects list of priorities.

Although it is sometimes very difficult to decide what the highest priority for a job is, the bottom line is that *there can only be ONE #1 priority,* and the owner of the facility must make the call.

Priority	Project Attribute
?	Price
?	Schedule
?	Quality

While we are discussing the price of the building, it would be a good idea to get a feel for the general cost of the commissioning services. Several generic magazine articles and technical papers have been written to justify the cost of this service, but few have gone so far as to suggest what the service is worth. That's why we were so appreciative when *Energy Engineering* published the following tables.

Table 7-2. Approximate Commissioning Costs for Design Team and Construction Team.

Team	Added Cost, Besides Direct Commissioning Costs
Typical Design Team Added Costs for Commissioning	25% of the Commissioning Design-Phase Costs
Typical Construction Team Added Costs for Commissioning	10-25% of the Commissioning Construction-Phase Costs

Source: *Energy Engineering,* **Vol. 104, No. 3, Page 19, "Tips for Applying Commissioning."**

One of the typical surprises within the commissioning decision is the fact that the services of both the design and construction teams are increased when the decision is made to add commissioning to the project. There are added responsibilities to coordinate specifications, attend meetings, and visit the jobsite. These added services should be integrated into the overall fee structure for the project. Based on what I've seen in the past, they will earn every dollar of it.

Then, there is the obvious expense of the commissioning agent, which can be generally presumed to fall within the ranges shown in Table 7-3.

Just to make sure you know what you are getting, every commissioning proposal should include a specific list of services you will get from their efforts *and* the services you should not expect from them.

Table 7-3. Approximate Commissioning Costs for Standard Building.

Scope Assumptions:		New Buildings = HVAC, Elec, and Controls Retro-commissioning= HVAC and Controls Only	
Project Cost	New, Detailed	New Basic	Renovation, Retro-Comm, all Basic Level
$ 0-50k	---		$5000 +5-7% of HVAC and controls replacement cost
$ 200k	---		12-14% of HVAC and controls replacement cost
$ 500k	---		10-12% of HVAC and controls replacement cost
$ 1M	---	2-4% of total construction cost	8-10% of HVAC and controls replacement cost
$ 2M	3-6% of total construction cost	1.5-3% of total construction cost	---
$ 5M	2-5% of total construction cost	1-2.5% of total construction cost	---
$ 10M	1-3% of total construction cost	0.5-1.5% of total construction cost	---

Source: *Energy Engineering*, Vol. 104, No. 3, Page 19, "Tips for Applying Commissioning."

Make no mistake, every commissioning job is different. Before the final task list is completed, the owner or CSO, architect, contractor, and sometimes even the building occupants have a hand in determining what is needed and not needed for their particular project. Based on the degree of technicality, the size of the project, the experience and confidence of the owner's staff, and the types of systems being installed, the services requested from the commissioning agent may vary greatly. It may be that only the HVAC and related control system needs to be commissioned. Then again, the owners staff may have no experience with electrical systems, so in that case, electrical system analysis should be included in the services.

On the other hand, one of the areas that often causes the owner to believe that commissioning is essential may not be critical at all. For example, many highly technical facilities will require the installation of equipment that absolutely must be installed correctly. This is certainly true for the more sophisticated medical and high tech facilities that are being constructed today. Because of the nature of the installed equipment, it is presumed that a commissioning agent must be employed to watch and facilitate during installation. Well, the truth is that no commissioning agent is so well versed in all things technical that he/she could be trusted to perform normal commissioning services on such equipment. These check-outs should be provided by the manufacturer's representative and not interfered with by those who don't understand the system or the installation processes required for those systems. Functional testing and final acceptance can only be provided by the experts in that field. So leave the commissioning agent out of it and put him/her to work with commissioning systems he/she understands and for which he/she is comfortable accepting some of the responsibility.

Here is a summary of the services and caveats we normally insert into our proposals.

What Commissioning **WILL DO**
- *Helps protect the client's investment.*

- *Reduces risk.*
- *Raises the "Bar of Quality."*
- *Provides early detection of coordination issues and systemic problems.*
- *Minimizes systemic problems (built-in problems.) The sooner a problem is detected, the less it costs to correct.*
- *Improved team coordination.*
- *Identifies and protects the project intent.*
- *Avoids typical pitfalls.*
- *Reduces change orders and construction-related problems.*
- *Reduces lost time from RFIs and other construction delays.*
- *Reduces warranty items.*
- *Allows new buildings to run smoother from the first day.*

WHAT COMMISSIONING **WILL NOT DO**

- *Take the place of owner involvement. Commissioning will likely fail without the owner's support.*
- *Commissioning agent does not serve as owner's financial agent, approve payment requests, direct contractors, attend all meetings, or produce meeting minutes.*
- *Commissioning will not replace the general contractor/construction manager's quality control activities during construction, nor will it replace the designer's punch lists.*
- *Testing and reviews will not be provided for 100 percent of the installed systems.*
- *No, the final project will not be free from all problems.*

The "?" in the "Owner Specified" column, in Table 7-4, represents the fact that the owner can modify the services of the commissioning agent to fit his/her project's need. No generic program is required, but the certification agencies do have a certain set of expectations that should be implemented by the commissioning agent selected for the job.

Well, the bell has rung and I have to conclude the chapter on commissioning for sustainability. But before I go, I have to admit that when this commissioning service first burst upon the MEP de-

Table 7-4. Typical Deliverables: New Building or Major Renovations.

Deliverable	Basic	Detailed	Owner Specified (1)
Document Design Intent and Narrative	•	•	?
Project Intent Workshop for Owner		•	?
Write Specification Additions to include Commissioning		•	?
Develop Commissioning Plan for MEP & Life/Safety	•	•	?
Review of Building Energy Modeling		•	?
Design Review – Schematic		•	?
Design Review – Design Development		•	?
Design Review – Final Documents, Before Construction Begins	•	•	?
Review Control Sequences of Operation	•	•	?
Submittal Review of MEP & Life/Safety Systems	•	•	?
Construction Observation	•	•	?
Attend Regular Meetings with Contractors (as needed)		•	?
Verify Indoor Air Quality Construction Practices	•	•	?
Test and Balance Verification	•	•	?
Construction Checklists		•	?
Creation of Functional Testing Programs for MEP & Life/Safety	•	•	?
Issue Resolution Participation	•	•	?
As-Built Drawing Verification	•	•	?
Control Sequence Verification	•	•	?
Training Plan Review	•	•	?
Oversee Training – (Provide Assistance)		•	?
Operations and Maintenance Manual Review	•	•	?
Near-End Warranty Review		•	?
Verify Delivery of Electronic Documentation		•	?
Re-Commissioning Manual		•	?
Final Report and Lesson Learned	•	•	?

Source: *Energy Engineering,* Vol. 104, No. 3, Page 19, "Tips for Applying Commissioning."

sign scene, I didn't support it. In fact, I deliberately placed myself on the other side of the issue, proclaiming that engineers and architects already provided these tasks and the added cost was simply going to make it more difficult to convince the owner/investor to sign the contracts and let us start the work. Well, it wasn't long before the reality of the data included within these pages finally convinced me that commissioning did deserve a place at the table and that the benefits it offers far outweigh the associated cost.

Much like energy management, commissioning has evolved into a necessary part of our world because the world just keeps on changing, and with the change comes a high tech arena where no responsible engineer or architect can claim expertise. As a result, we have to yield our old methodology and accept the new services that help provide the owner/investor with a facility that we can all be proud of.

Not only am I proud of the impact *commissioning* is now having on our industry, our firm has joined whole heartedly in its promotion and we are experiencing the true value that commissioning brings to our projects.

Chapter 8

Green Insurance

Bob Sansone

If you type "Green Insurance" into your browser, you're going to see that Google indicates nearly 100,000,000 results for this. If you had done this perhaps only 12 years ago, the results may have looked like this:

Results for: green insurance
    Your search—green insurance—did not match any documents.
Suggestions:
    Make sure all words are spelled correctly.
    Try different keywords.
    Try more general keywords.
    Try fewer keywords.

So what brought about this prolific explosion of sites devoted ostensibly to green insurance, and what is green insurance anyway?

Further, what is "green insurance" intended to address that "non-green" insurance does not? Perhaps more importantly, why should this information be meaningful to chief sustainability officers (CSOs). There's an old saying—"If you don't know where you're going, any road will take you there!" Similarly, not understanding the precepts of insurance is an almost sure-fire way to stumble into obtaining what you believed to be "green" coverage; only to find that what you bought sadly describes only the color of the money you spent. In other words, you may obtain a policy and labor under the false impression that you have covered the bases. When you critically need that coverage, you find out to your dismay that you secured something that may not respond as you expected it to.

It will help to start with a fundamental review of the concept of insurance and the insurance industry in general.

INSURANCE BASICS

Let's start with a roster of who the principle players are (sometimes referred to as the usual suspects!)

- **The Carrier** (aka insurance company)—This is typically an insurance company, which may be licensed and regulated by each state to provide certain classes of insurance coverage. The carrier or insurance company accepts the risk of loss at any location where it provides insurance and is compensated by the premiums it charges clients for the coverage it extends to them.

- **The Insurance Underwriter**—As part of an insurance firm, this is the final individual who determines whether a carrier will offer coverage to you. If it is a go, he/she decides what requested premium, deductible and other terms will be offered to accept the risk. For "green Insurance" is likely to be a specialist who has been certified or trained in this class of coverage and its unique characteristics.

- **The Risk Manager**—This is the individual in your own firm who is supposed to oversee the entire insurance and risk management process. This is also the individual who is responsible for selecting a broker to engage the carriers for insurance quotes and services. It is the duty of the risk manager to assess the array of coverage being assembled by the broker to also ensure that all risks are covered and that there is no gap in coverage. The risk manager also administers the program for the firm regarding loss prevention, risk mitigation, etc. If there are unique issues surrounding the insurance coverage of a sustainable building requiring "green insurance," the risk manager has to have the experience and training to recognize this and ensure these issue are incorporated into policy.

- **The Broker**—This is the individual who, as agent for your firm (authorized by the risk manager) is supposed to engage carriers in seeking the specific coverage for your firm, with broadest terms, at the least expensive price. The broker is not at risk for any loss. The broker for a client requiring green insurance needs to also have the training and experience necessary to recognize the unique issues involved in this specialty class of insurance. The broker takes his fee typically as a percentage of the premium you will be charged for insurance (this ranges, but is typically between 8-20 percent) and may be buried in the overall premium. The cost of engaging the broker may not be transparent or visible to the client. This fee is called the commission.

- **The Loss Prevention Specialist** (aka loss control specialist, risk engineer, loss consultant, etc., etc.) is an individual (typically employed by the at risk carrier) who is engaged to visit your location and write a risk evaluation or loss prevention report identifying risks that are observed (for which "green" insurance is being sought) along with recommendations to mitigate or obviate these.

- **The Treaty Reinsurer** is a fascinating entity and component, which is typically invisible to any client seeking green insurance (but is absolutely necessary for this class of insurance to even exist)! This critical component represents the majority of the capacity that a carrier is providing to a project. The truth is, in most instances; the majority of the capacity being offered by any carrier comes primarily from a group of other insurance companies (referred to as reinsurance companies).

They negotiate with these companies (mostly on a yearly basis) as part of their coverage treaty. Thus, when a firm represents that they can offer $200MM of green insurance capacity for your needs, what they are doing is extending their own capacity (their own money) for perhaps only 10-20 percent of this. The balance of the coverage being offered is actually behind them via a number of reinsurance treaty partners who have agreed to allow their own capacity to be extended under the banner of a single carrier for "green" insurance projects.

Insurance is typically constructed around two main drivers: the frequency and severity of losses that are contemplated in a risk class or occupancy. In this case, we have to understand what risks are contemplated within a building that has been designed or renovated to green status vs. one that has not. Or, where new technologies are being used in processes as opposed to a more traditional approach. Herein lies a fundamental issue that the chief sustainability officer (CSO) needs to understand: insurance companies effectively run their business as if they are driving a car through the rearview mirror. In essence, they look to past history in a specific occupancy (for example office buildings), which would be sufficient to establish a pattern of losses from which they can project future exposure (and therefore losses) they might encounter. In this manner they can establish deductibles, premiums and terms that they feel will allow them to make a profit.

Thus, if there is no past history, or it is new technology, the risk is viewed by the insurance industry as an exposure that does not merit extending insurance capacity. As we know, the proliferation of insurance offerings on the internet for green insurance products indicates that there must be a hefty appetite for insuring green operations with green insurance. So what happened?

With little exception, each policy of insurance takes the issue of frequency and severity and essentially structures their offering to policy holders around four main components: the deductible, the price or premium, the terms (the actual language which describes what is covered, excluded, sub-limited, what constitutes a claim, how a claim will be adjudicated, etc.) and the limit.

So let's assume a CSO, in concert with the risk manager, is seeking insurance for a building during the nascent emergence of green building design, technology, renovations, etc. Literally, there is no past history to drive the insurance model through for these new designs and technologies. What to do?

The effort now has to be one wherein the insurance industry, which loves to find opportunities to structure new product offerings for emerging markets (and then charge a premium), has to expand its search for historical support for why these green operations should merit coverage. To do this, they engage in an internal interrogatory to determine what is it about these new designs, technologies, etc. that can be extrapolated to existing designs which enjoy coverage.

THE UNDERWRITER'S STAPLES

Now let's introduce the three main staples of the insurance underwriter (be they "green" or otherwise) the uniform loss events: the normal loss event (NLE); the probable maximum loss (PML) and the maximum foreseeable loss (MFL). Here is a brief explanation of what they mean and why they are important to "green" insurance considerations.

Uniform Loss Events (ULEs)

ULEs are called by other names, but they are the backbone of putting a property policy together for any policy of insurance (be it "green" or otherwise). They are important for a variety of other reasons as well (carrier's treaty partners may limit them to what they can accept; they may limit an underwriter's authority to bind a particular type of risk, they are used to determine aggregation exposure, etc., etc.). They are also expressed typically as an expected dollar amount of physical damage and the expected dollar amount equivalent of time delay or business interruption for the location. Let us examine each and indicate how they may affect your "green" insurance policy:

- **The Normal Loss Event (NLE)**—The NLE effectively represents the highest frequency event for mechanical and electrical equipment breakdown in a sustainable or "green" occupancy when all protective devices and controls operate as expected and yield the lowest severity impact. To use an analogy, it would be similar to identifying the parts you expect to periodically replace on a car due to normal wear and tear, such as wipers, tires, brakes, belts and hoses, which are considered as normal maintenance and therefore not covered by insurance. The object is to then have the operator recognize the need at the inception of a problem with these (belt squealing, brakes vibrating, etc.) and the operator safely pulls the vehicle into the mechanic to have the offending part repaired or replaced. These are non-fortuitous events, as you expect them to fail at a certain point.

  Similarly, the objective of determining the NLE for green insurance is to define those events which are at the highest probability of requiring replacement or repair; you expect the building's maintenance personnel or the "green" control systems to signal the need for this repair or replacement. This results in the lowest financial impact for physical damage and business interruption/time element given your specific building and equipment array. Here is where you need a true "green" building expert to help define this, based on experience for your specific building and array of equipment. Most carriers will identify the NLEs present for the green building (every major piece of equipment will have its own unique NLE) and then select the highest of these to establish the NLE for that building. Or, in the case of multiple buildings, it may be the highest NLE for the entire account.

  Then the insurance company does something rather clever with this NLE: they try to establish this as the deductible for the "green" policy. By doing this, they to take themselves out of the non-fortuitous frequency of physical damage and business interruption loss for the account. As an example, for a certain single green office building of 20 stories, the NLE could be $150K and 30 days business interruption. This effectively means that the client will pay all physical damage losses below $150K, and the client will have to absorb any loss to revenues for a defined loss that is 30 days or less in duration.

- **The Probable Maximum Loss (PML)**—Having established the NLE for the "green" building or account, the carrier now has to determine what the most probable catastrophic event (aside from the natural catastrophe causes such as flood, earthquake, wind, etc.) that can happen to the facility given the specifics of the building they are reviewing. This event is referred to as the probable maximum loss (PML) and is the worst event that the carrier reasonably expects to encounter at the location. It would not be unusual for the insurance company to assume some adverse conditions would accompany this event. For example, they may assume that in the event of a fire, one of the fire suppression systems or sprinklers is not functioning, thus allowing the fire to spread further than expected.

  The PML for a green building is a function of their historical experience with similar green buildings and an assessment of what the building's insurance report indicates to them regarding type and design of unique or green equipment present (and the replacement value), layout for purposes of exposure to damage from mechanical breakdown, electrical failure or fire, building construction (combustibility, etc.), protective devices, quality of maintenance, human element issues, water supply, type and location of fire dept. or fire brigade, etc., etc. Effectively, they utilize the construction, occupancy, protection and exposure (COPE) information to determine potential losses (more about COPE below). The carrier will seek to determine the highest loss probable for this green building (or again for multiple buildings for the account) and establish this as the expected PML for the building/account. And again, not unlike the NLE, the carrier does something fairly clever with this PML: they can use this to help establish what they want to charge for this green risk. Thus, the higher the probable maximum loss, the higher the premium is likely to be.

- **The Maximum Foreseeable Loss (MFL)**—The MFL is somewhat an exercise in insurance engineering imagination. The insurance loss control specialist is typically asked, having defined the NLE and PML for the building or account, to extrapolate the very worst situation that could occur beyond the PML to bring about the most damage or loss conceivable (ultimate severity). In such an assessment, the following conditions are typically stipulated: all protective systems and devices are out of service and the fire department never arrives or arrives too late to make a difference; there is nothing to halt the spread of fire vertically; and the spread of fire to adjacent structures is assumed (unless they are a defined distance away) with the presence of continuity of combustible material which would allow the fire to spread. It would not be unusual for the efforts of the risk engineer in such an endeavor to conclude that the entire building could be lost. This situation is known as MFL = total installed value (TIV). Once again, the insurance company utilizes this value to help guide their underwriting decisions on the account and may use this value to help construct the limits on the policy they will offer. It is the lowest frequency and highest severity event imaginable.

Thus, with the establishment of these ULEs, the carrier can, among other things, establish their deductible, establish their pricing and create the limits they are willing to write the policy for. The insurance fundamentals for "green " and non-green insurance in these matters are very similar.

"COPING" WITH INSURANCE

Let's call this a search for analogies. So, if we have a regular (non-green) building, part of what an insurance company reviews before extending any coverage is called COPE (Construction, Occupancy, Protection, and Exposure). Let's review COPE before I deal with the analogy issue for green insurance.

Construction

This essentially allows the insurance company the opportunity to determine the structure itself; whether it is made of combustible material, non combustible material, age of the structure, roof characteristics (flat roof, peaked roof, hurricane protected roof, etc.), and anything else that helps them to see if this type of construction (based upon their loss history) presents a building which suffers a higher incidence of loss or not. This determines not only if coverage will be offered, but will also help to determine deductible requirements and premiums. Additional considerations are the number of stories the structure has as well as the exterior skin of the building, window design, interior finish, etc.

Occupancy

A very important component in assessing structures for insurance involves determining what is the primary activity being engaged in this building. Clearly, if the building (or occupancy) is involved with manufacturing explosives, then the risk of loss is dramatically different from an occupancy involved with retail banking or general office operations. Effectively, the occupancy describes the hazard classification of the building based upon the primary functions being performed inside.

Protection

For nearly all occupancies, the presence of appropriately designed protective devices (mechanical and electrical), fire suppression and/or sprinkler protection, alarms, location of fire department, etc., are all critical insurance considerations. Along with this, further examination of the reliability of the water supply feeding hydrants and sprinklers becomes an important issue subject to review.

Exposure

This deals with issues such as neighboring occupancies, which may expose the target occupancy. For example, a 4-story steel reinforced concrete structure, with essentially non-combustible construction, a non-combustible roofing system, with full sprinkler protection and housing a banking operation, which sits adjacent to a gasoline refinery, is exposed to losses from events occurring at the adjacent facility more so than from the bank itself. The same holds true for geographic issues such as exposures to flood, quake, and wind.

So, let's take a time machine back to 1993 when the precursor to green building and design was in all likelihood initiated in part by the Natural Resources Defense Council (NRDC) which initiated Leadership in Energy and Environmental Design (LEED). From this effort, a Green Building Rating System, developed by the U.S. Green Building Council (USGBC) emerged. This system provides a suite of standards for the environmentally sustainable design, construction and operation of buildings. However, the early parameters for what constituted a green building were only in the nascent phase.

Now imagine yourself as an insurance company underwriter presented with a request for coverage of a newly designed and built (or renovated) "green" building back then. Your first question has to be: What does this mean? Is the building less apt to have a claim or incident, has the COPE profile changed from what I am used to seeing, has the replacement value changed as a result of these new green standards and additions? Is the equipment (mechanical and electrical) more sophisticated, more apt to fail, and therefore more apt to result in a claim, is the fire exposure the same/less/greater, etc., etc., etc.?

So, you look in your rear-view mirror and find a great big

blank. No history, no loss data from which to extrapolate what a proper premium should be, no clue as to what a reasonable deductible should be and no idea of what limits you need to place on the policy (much less special terms). What to do?

Then you remember that obscure course you took in math (math....of all things!) about mathematical problem solving through analogies, and you start to ponder whether you can find operations, which you cover already, that may be sufficiently similar to this newly titled "green" building. Hmmmm, an analog to this "green" building?

So, you start to look at the equipment that helps the building function in a more green manner, such as:

- HVAC;
- electrical
- roofing
- solar panels
- high efficiency glass windows
- use of renewable components
- new insulation products

What you discover, perhaps incrementally, is that you indeed have analogs for these various items, for which you already are providing insurance coverage.

For example, let's say you are faced with determining the insurance implications of a green building (with its state of the art energy efficient components, materials, etc.) but you can't get a firm handle on the risks associated with the construction aspects of this (the C in COPE), and you set about looking at other buildings for which you have recently provided coverage.

Lo and behold, you find that many of these have indeed incorporated much of the construction material and techniques, which made them more efficient, but did not cause them at the time to categorize themselves as green building construction.

The increase in the application of these green construction materials in this new building is simply an expansion of what you have already covered in other non "green" buildings. Thus, your analogy is created and allows you to create a path to accepting coverage of this aspect of the risk. You do the same with the other aspects of the building and find similar analogies. So, with analogs in hand, you can commence to bridge the gap and underwrite the property for coverage.

TURNING GREEN

Let's now look at what insurances may be affected when seeking so-called "Green Insurance." Remember, the greater question for the insurance carrier, who offers "green" insurance, is: does a "green" design or building have a different propensity for losses than traditional non "green" designs and buildings (thus impacting frequency and severity)? If so, is it less or more; and what is required to normalize the risk if it is higher?

Starting with the design and construction of the facility, a class of insurance typically referred to as construction all risk (CAR) or engineering all risk (EAR) will be required by the financing entity. This coverage has a number of components, but generally responds to losses during the construction, start up and commissioning. The important issue here is that a CSO recognizes that he/she is designing and building a structure that has sustainable, eco-friendly, energy efficiency components designed into it (passive or active), which renders it "green." Insuring such a structure has implications for the client and the insurance carrier. For example, on a square footage basis, is the building more expensive to design and build (and therefore repair or replace) than a non-green building? The cost adder to design and build a green commercial project is currently estimated at 2 percent, but it is expected that this cost differential will dissipate and reverse to a cost advantage in the future. Does the introduction of green technology raise or lower the risk profile for equipment failure or fire potential? All of this effects the policy of insurance, including but not limited to premiums and deductibles.

In CAR or EAR coverage, it is important to understand the insurance company is accepting the risk with the tacit understanding (hope) that the project will be designed, built and commercially accepted without any losses; or losses which in aggregate are less than the total premium. It is also important to understand that at the inception of any CAR/EAR coverage, there is effectively nothing of insurable value at risk. The most pronounced risk (and therefore exposure for the insurance company) happens as the project nears completion and is preparing for commercial acceptance via startup, testing, commissioning, etc. It is at this time that all, or most, equipment has been installed, all tenant finish is complete, and the building is at, or near completion and represents a project with its full value in place. Insurance companies who know what they are doing can make a tidy profit on insuring such projects as they do not typically experience losses and they represent an elegant path to insuring the operation once the project is complete. But not always.

A tragic example of this (even though it was not a "green" project) recently occurred in Middletown, CT, and is a stark reminder of the fact that risk is ever present, whether it is associated with standard or green insurance. In a series of articles by Josh Kovner from the *Hartford Courant*, I have selected the following extract as a brutal reminder that in the insurance industry, as well as in the construction of projects, risk never sleeps and incidents can occur in the blink of an eye just when a project is preparing to be completed and go commercial. The tragic result in this case was that six people were killed and scores were injured. The EAR/CAR coverage provided by several insurance firms will clearly be taking a major financial hit as a result of this incident:

Middletown Building Official Declares 6 Major Buildings At Kleen Energy Power Plant Unsafe
By JOSH KOVNER
The Hartford Courant
March 11, 2010
MIDDLETOWN—

The owners of the Kleen Energy power plant, extensively damaged in a Feb. 7 natural gas explosion that killed six people, face a substantial rebuilding project. Six major buildings at the site have been declared unsafe by Middletown's building official, John Parker. These include the plant's central core—the power block building that houses the three turbines, the generators and the pair of 220-foot high stacks that make up the heat-recovery system. "They'll either have to knock them down or make them safe," said Parker. He said the contractor will have to obtain demolition and building permits. ...The majority owner is Energy Investor Funds of Boston. Working with investment banker Goldman Sachs, EIF amassed nearly $1 billion in financing for the plant. O&G, in brief statements after the blast, said that it intends to rebuild the 620-megawatt plant, which was designed to burn both natural gas and oil. ...No single state agency has overall jurisdiction to monitor such a project, and investigators said that safety protocols appear to have been violated during the Feb. 7 purging.

It is important to note that errors or oversights, which are either inadvertently designed into a project or installed in error during construction, often times do not manifest themselves until the project has been completed and is in commercial operation. Thus, the CSO needs to ensure that design and construction of the project employs technically proven and accepted best practices for green design and construction. Remember, the underwriter will be seeking out these same issues to determine if the project merits coverage consideration. Best practices will be a key for all stakeholders through all phases (design, build, operation, maintenance).

It is also not unusual to have a period of time wherein maintenance is covered during the initial operation (referred to as completed works coverage). The EAR/CAR coverage also responds to natural disasters or elemental perils (quake, flood, windstorm, etc.) in addition to fire, mechanical breakdown and project delay for a covered event prior to commercial acceptance. The question has to be asked therefore: is there any unique characteristic of this green design that raises the risk profile for the coverage being sought? For example, if the building employs solar panels, are these more susceptible to loss or damage from wind or hail?

EAR/CAR coverage will not, however, respond to a failure of the design to work effectively or efficiently. Thus, if the green design parameters indicate that the building using the green materials and technology will only require X Btus to maintain an ambient temperature of Y, or use Z less kW than a non-green building in defined spaces, and it fails to do so, the failure is not a covered event under insurance. Such coverage would be referred to as performance guarantee or efficacy coverage and it is unlikely a CSO, or risk manager, will find it available in the market today. Thus, if you are seeking a product from the insurance industry that insures efficiency of design, you are not likely to find it. Issues like this are typically covered under contractual arrangements defining testing and acceptance parameters to prove the designs efficacy. The current state of the industry indicates that attracting EAR/CAR insurance capacity for green projects should not be a problem.

The EAR/CAR coverage typically terminates upon commercial acceptance of the project. At this point, a new class of "green" coverage for the operation is required, and these are referred to as the property and casualty covers (P&C). Contained within these policies are coverages for equipment breakdown, fire, elemental peril, pollution and liability.

Structuring liability insurance provisions for green projects or buildings will likely be no different than non-green ones (with the exception of any "green" component or material being discovered to have negative health consequences for humans while in the building. An example would be the issues faced if the construction industry discovers that certain insulation that was considered green and eco-friendly, was actually degrading over time and putting forth toxic gas or particles.).

The need will exist for the "green" insurance coverage to also respond to time element disruptions (or business interruption as it is commonly referred to) during normal operations. To accomplish this, there must exist either an agreed upon process wherein parties can stipulate to the value of such a delay, or a process whereby actual losses sustained are calculated. The key is to determine whether the "green" aspects of such a loss, disruption or delay is greater than, equal to, or less than other non-green insurance coverage for a similar project or buildings

To illustrate, let us speculate that a non-green office building is shut down for 10 days (ignoring deductibles for the moment) due to a catastrophic failure of the HVAC system.

The building presumably has an impact to the rental revenue stream, and the rent for tenants in a "green" building may be greater than a non-green building. Thus, the time element exposure to an insurance company may be more pronounced and, therefore, cost more.

Similarly, the HVAC for a "green" building may have a larger capital exposure for replacement than a non-green building. Thus the replacement cost for such a building may have premiums or deductibles somewhat greater based on repair or replacement costs. The valuation of property is one of the biggest challenges for the insurance industry as often times they are rating the account for the purpose of calculating premium on the cost or estimate of the property.

To effectively construct coverage for such a time element event, there must be an advanced agreement upon the daily or hourly value of such an interruption. Establishing this in the covenant of insurance obviates problems of adjudicating the impact of a business interruption or time element type of loss and affords all parties more certainty in the process.

Once identified and agreed to, insurance companies can price such a time element risk into their premium, establish deductibles to try to reach a fair balance of what a client should pay before insurance inceptions, and clients can structure their operation consistent with what their financing partner is willing to accept.

It is well to keep in mind that insurance costs for clients should be lower as the risk is evaluated to be lower. If there is data supporting increased productivity, fewer absentee/sick days, fewer losses, etc.; then, there should be reduced insurance premium for such a certified building. In fact, Firemen's Fund open market approach validates this approach. In an apple to apple building comparison, the building certified as green will enjoy a reduced premium over the non-green building from their underwriting dept. The same holds true for their clients, who build or renovate to green standards. As to the potential impact for health insurance costs, if the case can be put forth and supported that employees working in green buildings are overall healthier and thus have a smaller impact on health costs; then, it would follow that such costs for this class of insurance could also be reduced.

FINANCING

Let's turn our attention to the cousin of "green" insurance for a moment: "green" financing. A CSO contemplating "green" insurance would be well advised to have a good long collegial discussion with the proposed financiers. Many of the same risk considerations for green insurance are going to be reviewed by the entity that is putting forth the risk capital to design or build the project. And trust me when I tell you they are as risk averse as anyone you are going to find these days.

In their article, "Two of the Three Little Pigs Would Have Trouble Getting a Loan; Odd Homes Built of Tires and Trash Lure Environmentalists, Turn Off Bankers," Anton Troianovski and Nick Timiraos point out the risks of over-reaching on the environmental friendly design-side for what would be considered commercially viable for the financing industry for residential homes. It is a solid rule in the industry: if you have problems financing the project, you will have bigger problems attracting capacity to insure it! The financing lessons of the residential "green" and sustainable projects should not be lost on the commercial scale project developers and those seeking green insurance. Chief sustainability officers take note!

Ensure that the sustainable and green components of your project are technically proven, capable of being replicated, are readily replaceable, and most importantly, are accepted as proven designs. Remember, insurance is predicated on repairing and/or replacing something; and if the design or technical component is considered as one-off or experimental, the CSO and risk manager, will not find the insurance capacity they seek. CSOs are urged to have a dialogue with the bankers and see what they are thinking! By the way, bankers will start the process of structuring the necessary insurance program (minimum deductible, time element component coverage, etc., etc.) so they can avoid the risk!

The Broker

Suffice it to say that "green" insurance for commercial scale projects will typically not be underwritten without the help and assistance of a broker or agent. Many brokerage firms profess to have specialty services available for specific risk segments, insurance lines of coverage and industries such as casualty, directors and officers insurance, energy, medical, workman's comp, etc.

A measure of how far "green" insurance has come as a recognized segment; CSOs should look at brokerage firms professing to have such expertise and specialization. Literature on the topic can lead to good sources of information and potential brokers. One good example is the green insurance article by Andrea Ortega-Wells, who wrote "Under Construction: The Green Property Insurance Market." She wrote about the efficacy of such insurance, and her sentiments in that article include the comment, "A recent survey by McGraw-Hill Construction projected the near-term market growth in green construction for the following building sectors: Education, 65 percent; Government, 62 percent; Institutional, 54 percent; Office, 58 percent; Health care, 46 percent; Residential, 32 percent; Hospitality, 22 percent; and Retail, 20 percent. Owners and developers of commercial and institutional properties in North America are advancing green development through state-of-the-art tools, design techniques, advanced green products and creative use of financial and regulatory incentives."

Having the benefit of Ms. Ortega-Wells perspective on "green" insurance, it is only fair that we hear from the carrier side of this equation on what "green" insurance means to them. Fireman's Fund was generous enough to allow me the time to discuss green insurance with Steven Bushnell, who heads up their "green" insurance practice. Fireman's Fund was apparently the first carrier to offer commercial and residential green insurance catering to certified green buildings.

The company was able to do this, according to Bushnell, as a result of its review of past history (remember the rear view mirror and analog assessment?). The company discovered that its losses for commercial buildings over the past 10 years had, as primary and secondary causes of loss, electrical and plumbing matters. In researching green buildings and design, Fireman's Fund determined that the risk of loss from these exposures were actually lower for "certified" green buildings; and as such, it determined construction and design merited a discount.

This is not to say that the company simply accepts the building as self-certified. In fact, it looked to LEED standards from the U.S. Green Building Council. It also is understandably concerned with the introduction of new materials meeting sustainability standards and parameters. The company sees a tremendous market, according to Bushnell, as there are over 5 million commercial buildings in the U.S. that are not certified "green." It offers the opportunity to not only provide the building (if upgraded to green) a policy advantage, but also can do this on a tenant basis for their insureds who occupy and rent rather than own their space.

Fireman's Fund estimates that green renovation or upgrade to existing buildings afford the owner with a 20-30 percent reduction advantage in building operating costs alone. Bushnell also points to other statistics for green buildings which (among other issues) infer a reduced absenteeism rate for employees in green buildings, better test scores in green schools, higher resale value of green buildings, increase in the square footage lease value of a green buildings, etc.

It was also fascinating to hear about the concerns that Fireman's Fund has regarding green building applications when they are extending coverage. For example, one hallmark of a sustainable energy building has to do with their commitment to recycling. While the recycling aspect is encouraged, the green insurance aspects are affected from the exposure to accumulation of combustible material in loading docks and other areas that could provide an initiation source or path for fire to travel, which would otherwise not be present. To address this, they ensure that recycling is ideally handled off premises.

Another reported aspect of green insurance that differs from standard insurance is the aspect of covering fully (without sublimit) roofing structures, which may employ vegetative roof systems for efficient cooling and underground water retention for water conservation (cisterns, etc. for cooling and landscaping).

The bottom line is that Fireman's Fund initiated this commercial coverage in 2005 and, in effect, it established this market. This was followed by its introduction of green insurance for residential in 2008. Thus, it has been the leading edge for the green insurance movement and is fully committed to it. To do this, the company has answered the questions rather affirmatively which were asked earlier in this chapter:

- Is a green building more or less apt to experience a loss? Their view—Less.

- Does a green building require a set of standard certifications to classify it to assure risk assessment (and therefore a policy) is structured on a set of reliable standards? Yes and this exists through LEED.

- Is the replacement value of a green building more or less expensive than a standard building? More by around 2 percent; however, because of the lowered risk profile, the premium applied may be less!

In conclusion, the time appears to have arrived for a mature, green insurance market. What it likely requires now are informed buyers of this product; i.e., sustainability managers who understand the various issues and options that impinge upon green insurance. It also needs an economy that affords the marketplace the opportunity to weigh the prospect of going green as well as the financial benefit.

Putting aside your political philosophy for just a minute, one might consider:

- the current financial need to improve our economy by efficiency measures and reducing costs;

- the need for a common sense approach to ensure we move toward energy independence; and

- the desire to responsibly mitigate our impact on the environment with more efficient use of our energy.

In addition, consider that green insurance acts as both a more economical alternative to "standard" insurance (in many applications) as well as a compelling incentive/motivating factor in helping clients shift their emphasis toward getting their buildings certified as green. Taken together, this should, assist in furthering the sustainability movement.

Chapter 9

Sustainability and LEED Green Buildings

Nick Stecky

As of this writing, Leadership in Energy and Environment Design (LEED) has been with us for about 10 years. The marketplace has had considerable experience with LEED, New Construction (NC), and is developing experience with LEED Existing Buildings (EB), and Operations & Maintenance (O&M). Over the past decade, there has been plenty of enthusiastic acceptance of LEED, but there has also been some criticism, especially as it relates to the actual energy efficiency performance of a LEED building.

Unfortunately, there is a bit of confusion and unrealistic expectation of LEED NC as NC focuses on design and construction, but not on operations. During the design charettes, energy efficiency and all other credits are prioritized. It is possible to relegate energy efficiency to a low priority, if the owner wishes. The building could still achieve a LEED certification, even earning the minimum of energy points. NC energy efficiency credit points are earned by performing a computerized building simulation of the building before it has actually been operated for a reasonable period of time, such as one year or more. Too often, it is an educated guess—only an indicator of actual performance.

If the interest is on actual energy performance, then we need to look to LEED EB O&M. In this rating system, energy points are earned by collecting at least a year or more real world operating data and performing an EnergyStar portfolio analysis on the building and systems.

It helps to compare a newly completed and documented NC project to a new college graduate. They have gone to the best school, took the best courses and graduated with honors. They have a great start! But how well will they perform in the actual workplace? We know from experience that good schools and grades do not necessarily deliver an outstanding career. It depends upon how well that career is managed and the effort put into it. The same applies for LEED. It is EB O&M, which will maximize the chances of an outstanding lifecycle performance of a building and the degree to which it satisfies our sustainability needs..

THE RISE OF ENVIRONMENTAL CONCERNS

Every chief sustainability officer (CSO) needs a sense of where LEED, and more importantly, the sustainability movement began. The publication of Rachel Carson's *Silent Spring* in 1962 alerted the general public to the dangers of pesticides, in particular the dangers to humans. This helped precipitate the rise of an environmental movement, politics, legislation and regulation in the United States during the sixties and seventies. New laws were passed to protect the environment. These included:

- The National Environmental Policy Act of 1969—this created the U.S. Environmental Protection Agency;

- The Clean Air Act was passed in 1970. This greatly expanded the protection of two previous laws, the Air Pollution Control Act of 1955 and the first Clean Air Act of 1963;

- The Water Pollution Control Act of 1972;

- The Endangered Species Act of 1973; as well as

- The formation of the U.S. Department of Energy in the late seventies.

As these laws were passed, there came a tension, an apparent conflict, between the need to preserve the environment and the need to grow and expand the economy and jobs. Environmentalists began to be seen as opponents of growth and industry. There appeared to be a contradiction between business and protection of the environment. Environmentalism began to be seen as just another "special interest" group which simply added cost to running a business with very little added value. Sustainability began to build a bridge between the two. This movement has shown that you can have both a clean environment and strong business climate.

During the eighties, in reaction to high energy costs, inadequate energy supplies, pollution control and other environmental factors, a new approach to designing, building and operating buildings began to develop. While buildings and occupants were suffering through difficulties as previously described, the nineties gave rise to the sustainability movement. It was in 1999 that the book *Natural Capitalism* by P. Hawken, A. Lovins, and L.H. Lovins was released. The authors presented the connection between economics and environmentalism, and argued that they are mutually supportive, not mutually exclusive. As *Natural Capitalism* points out in its title, nature itself is capital. For example, what is the value of a clean lake? It is the relatively high property values of houses along a pristine lake. It is drinking water that needs less expensive water treatment before being potable. It is the recreational value and income, through swimming, water skiing, boating, fishing and the rest. If the lake were degraded through pollution, property values would fall, drinking water treatment costs would rise, and lake use revenues would go down. Seems simple enough, but it has been mostly ignored until now.

Historically, the environmental movement consisted primarily of regulation, legislation, mandates and fines. But a new form of sustainable green revolution has been emerging. This one may succeed where traditional legislative environmentalism has had limited success. This new form takes a larger view, taking economic, community, and technological considerations into account as well as environmental concern. The idea of sustainability recognizes the need to consider cost-benefit analysis in evaluating and/or promoting various programs.

SUSTAINABILITY GIVES RISE TO
THE GREEN BUILDING MOVEMENT

The late eighties and early nineties were a crucial developmental period for the green design movement. Leaders of green design included William McDonough, Paul Hawken, John Picard, Bill Browning, and David Gottfried, who later went on to be one of the co-founders of the United States Green Building Council (USGBC). The movement acknowledged that buildings represent a very significant usage of resources, land, and energy. It further stressed that improvements to the ways in which we design, construct, operate, and decommission buildings could make significant contributions to improvement of the environment and overall sustainability.

Today, owners, occupants, and communities are beginning to hold buildings to higher standards. Industry leaders are responding by creating physical assets that save energy and resources, and are more satisfying and productive. Building owners, who want the greatest return on investment, can take a path that is green, both economically and environmentally. How are they doing it? They are doing it through integrated solutions for the design, construction, maintenance, and operations, as well as the ultimate disposal of a building.

An integrated, holistic, whole building approach, may mean life cycle costs over the life of the asset will be lower. By designing, building, and operating in an integrated way, owners can expect high performance buildings that offer:

- More efficient use of energy alternatives;

- Increased efficiencies of systems and resources;

- Quality indoor environments that are healthy, secure, pleasing and more productive for operators and occupants;

- Optimal economic and environmental performance;

- Wise use of building sites, assets and materials;

- Landscaping, material use and recycling efforts inspired by the natural environment; and

- Lessened human impact upon the natural environment.

GREEN BUILDING RATING SYSTEMS

There are a number of green rating systems that have been developed that the CSO should be aware of. One of which is LEED. Each has its own merits and depending upon what criteria one desires for his/her "green" building will determine the choice of rating system. For example, the Energy Independence and Security Act of 2007 (EISA) requires energy efficiency and sustainable design standards for all new federal buildings and includes a certification requirement. EISA does not specify which rating tool to use, simply points out required criteria. When the federal government's General Services Administration (GSA) was looking to comply with EISA, they utilized a 2006 report from Pacific Northwest National Lab (PNNL). This report entitled, *Sustainable Building Rating Systems: Summary, PNNL-15858 prepared by KM Fowler, and EM Rauch.* The report identified five rating systems as having the greatest potential of addressing GSA needs. The PNNL report summarized and reviewed each of the five rating systems, but did not make a recommendation on a preferred system. The rating systems studied and reviewed by PNNL in the report were:

- CASBEE, Comprehensive Assessment System for Building Environmental Efficiency

- GB Tool

- BREEAM, Building Research Establishment's Environmental Assessment Method

- Green Globes, US

- LEED, Leadership in Energy and Environmental Design by the USGBC

GSA evaluated these five ratings systems based upon four criteria: applicability, stability, objectivity, and availability. The agency decided the LEED rating system was the best fit for its needs. Other federal agencies have also adopted LEED as their rating system. In general, LEED has become the most widely accepted green building rating system in North America.

Currently, federal agencies, such as DOE, NASA, DOD, etc., use LEED and require a minimum of a Silver rating. They also require a minimum energy efficiency of 30 percent better than required by the ASHRAE 90.1-2007 Energy Standard for Buildings Except Low-Rise Residential Buildings.

UNITES STATES GREEN BUILDING COUNCIL

In 1993, David Gottfried, a developer, Mike Italiano, an environmental attorney, and Rick Fedrizzi of Carrier Corporation got together to form the United States Green Building Council, USGBC. They had become concerned about the fragmentation of the building industry and the absence of consensus related to sustainable design.

The USGBC was initially formed to create an educational organization that would bring building professionals together to promote sustainable design. In 1997, the USGBC was awarded a $200,000 grant from the U.S. Department of Energy, and the organization was off and running. The USGBC has made a large impact on the design and building industry. There are more than 5,500 members consisting of individuals, leading corporations, governmental entities, universities, educational entities, consultants, product manufacturers, trade associations, and more.

The mission of this unprecedented coalition is clearly to accelerate the adoption of green building practices, technologies, policies, and standards. The organization is a committee-based organization endeavoring to move the green building industry forward with market-based solutions. Another vital function of the council is linking industry and government. The council has formed effective relationships and priority programs with key federal agencies, including the U.S. DOE, EPA, and GSA.

The Council (USGBC.org) is viewed as a balanced consensus coalition representing every sector of the building industry. It spent five years developing, testing, and refining the LEED Green Building Rating System. When adopted from the start of a project, LEED NC helps facilitate integration throughout the design, construction, and operation of buildings.

LEED OVERVIEW

The LEED Green Building Rating System™ is a voluntary, consensus-based, market-driven building rating system based on existing proven technology. It evaluates environmental performance from an integrated or "whole building" perspective over a building's life cycle, providing a definitive standard for what constitutes a "green building."

LEED is an assessment system that incorporates third party verification and is designed for rating new and existing commercial, institutional, and high-rise residential buildings. It is a feature-oriented system where credits, also called points, are earned for satisfying each criterion. Different levels of green building certification are awarded based on the total credits earned. The system is designed to be comprehensive in scope, yet simple in operation.

LEED Green Buildings

Green buildings are designed and constructed in accordance with practices that significantly reduce or eliminate the negative impact on the environment and its occupants. This includes design, construction, operations, and ultimately, demolition. The LEED Rating System consists of five fundamental categories:

- Sustainable site planning;
- Safeguarding water and water efficiency;
- Energy efficiency and renewable energy;

- Conservation of materials and resources; and
- Indoor environmental quality.

LEED strives to encompass a wide band of sustainability that includes:
- Society and community—recognizes that buildings exist to serve the needs of individuals and the community, but that they must also minimize their impacts;

- Environment—again striving to minimize negative impacts on the environment;

- Economics—recognizes that the adoption of these sustainability initiatives by business will be a function of the economic benefits that can be delivered by green buildings; and

- Energy—recognizes that energy plays a key role in building operating costs as well as a sustainable energy future.

Benefits of Leed Buildings

At the onset, LEED was created to standardize the concept of building green, to offer the building industry a universal program that provided concrete guidelines for the design and construction of sustainable buildings for a livable future. As such, it is firmly rooted in the conservation of our world's resources. Each credit point awarded through the rating system reduces our demand, as well as our footprint, upon the natural environment.

Conventional wisdom says that the construction of an environmentally friendly, energy efficient building brings with it a substantial price tag and extended timetables. The CSO needs to know that this need not be the case. Breakthroughs in building materials, operating systems and integrated technologies have made building green not only a timely, cost effective alternative, but a preferred method of construction among the nation's leading professionals.

The USGBC summarizes it well, stating, "Smart business people recognize that high performance green buildings produce more than just a cleaner, healthier environment. They also positively impact the bottom line. Benefits include: better use of building materials, significant operational savings, and increased workplace productivity."

In many instances, green alternatives to conventional building methods are less expensive to purchase and install. An even larger number provide tremendous operational savings. The US-GBC and LEED offer the building industry a fiscally sound platform on which to build their case for whole-building design and construction. Green buildings can show a positive return on investment for owners and builders.

Economic benefits also can include:

- Improved occupant performance—employee productivity rises, students' grades improve. (California schools have a number of analyses that show children in high performance facilities have improved test scores.)

- Absenteeism is reduced.

- Retail stores have observed measurable improvements in stores with daylighting.

Building green also enhances asset value. According to organizations such as the International Facilities Management Association (IFMA) and the Building Owners and Managers Association (BOMA), the asset value of a property rises at a rate of ten times the value of the operational savings. For example, if green building efficiency reduces operating costs by $1/sq ft per year, the asset value of that property rises by ten times that amount, or $10. It pays to be efficient.

In addition to the environmental and economic benefits of building green, there are also significant societal benefits. These include increased productivity, a healthy work environment, comfort, and satisfaction, to name a few. The good news for CSOs about green buildings is that they can then be leveraged with local and trade media through press releases, ceremonies, or events. By calling attention to the building and its certification status, owners speak volumes about themselves. The USGBC has stated: "Like a strong prospectus, building green sends the right message about a company or organization: it's well run, responsible, and committed to the future."

Benefits to the Architectural and Engineering Community

All too often, market forces drive a building design team to focus on minimum first cost regardless of what the overall life cycle costs might be. The end result is that all elements of society from owner to occupants, and to the community, end up with a building which is less than what it could have, or should have, been.

Use of the LEED process prompts the forming of an integrated design team from day one, and promotes the formation of a creative solution to the particular buildings needs being planned. As the CSO gathers this creative roundtable of equals, the team is able to maximize use of the collective wisdom of the members and develop a design concept that can be both green and economic. For example, increasing the use of natural light to displace some artificial light can result in a reduced load on air conditioning systems. These AC systems can then be downsized, resulting in reduced equipment first costs and reduction in electricity use, both through the artificial lighting reduction and reduced cooling loads. Small savings multiply and reverberate throughout the design, ultimately having significant impacts on overall first costs and operating costs.

When building "green," the sum can be larger than the individual parts when all components are integrated into a single unified system. Integration draws upon every aspect of the building to realize efficiencies, cost savings, and continuous returns on investment.

The most significant benefit to the A&E design community is that LEED promotes and rewards creative solutions. Sustainable and green designs are not yet commodity skills that all firms can lay claim to. For those A&E firms seeking to provide more value to their clients, LEED is a way to achieve this. LEED™ can become a standard for design excellence, and provides an A&E with a brand differentiator. LEED can become an outstanding competitive edge for firms seeking a leadership position of excellence.

LEED Acceptance

There's a groundswell of acceptance taking place. Corporations are requiring LEED in new construction, the educational sector from K-12 to higher ed are requiring it, as well as state governments and the federal sector.

When LEED was first introduced several years ago, there was reluctance to accept it. It represented a new way of looking at the design, construction, operations, and disposal of facilities. There were many questions about higher first costs, paybacks, overall benefits, ability of the design community to deliver, doubts about the technologies involved, etc. The perception was that green buildings were too futuristic, unattainable, and required too many tradeoffs to be able to deliver a practical, efficient facility where people could live, work, and play.

CSOs will find that some elements of these doubts remain, especially as to the financial benefits. However, as we gain experience with green buildings, we are developing the experience and the data needed to resolve these doubts. Green is becoming more cost neutral. In an article titled "The Costs and Financial Benefits of High Performance Buildings," Greg Kats of Capital E analyzed 40 California LEED buildings for the "cost premium" of the various levels of LEED certification. The study consisted of 32 office buildings and eight schools. Katz found that the eight LEED-certified buildings cost an average of 0.7 percent more, the twenty-one LEED silver buildings cost an average 1.9 percent more, the nine gold buildings cost average of 2.2 percent more, and the two platinum buildings cost an average of 6.8 percent more.

What is LEED?

LEED is generally considered to be:

- A rating system to measure the "greenness" of a facility based on the five fundamental aspects of green buildings: sustainable sites, water efficiency, energy & atmosphere, materials & resources and indoor environmental quality.

• A performance-based and prescriptive-based criteria.

• Guidelines which focus on whole building system instead of the components.

• Life cycle based, not first cost.

• A catalyst to promote architectural and engineering innovation; i.e., through LEED credits.

• A tool to provide an independent third-party verification procedure to ensure quality and compliance.

The LEED family of rating systems is growing in number and evolving to better address the needs of the marketplace. When the first LEED NC rating system was released in 1999, the USGBC hoped to be able to have an impact on the built environment. They have succeeded beyond their wildest dreams. Symbolic of that overwhelming acceptance is the marketplace demand for more tailored LEED rating system products for specific markets, such as for retail, schools, labs. In addition to the LEED NC and the LEED EB O&M, more tailored rating systems include:

LEED for K-12 Schools
LEED CI for Commercial Interiors
LEED CS for Core & Shell
LEED H for Homes
LEED for Campus

Other LEED programs are being developed in response to marketplace demand such as Labs 21.

The latest version consists of the following:

• Structured around the familiar 5 elements of site, water, energy, materials & resources and indoor environmental quality.

• Not a teardown and rebuild of LEED 2.2 but a reorganization and reweighting of credits.

- More emphasis on credits that reflect climate change, energy efficiency and cost efficiency.

- Because some environmental issues are unique to a locale, regional councils have identified zones for regional priority credits, which can earn an additional 4 points.

- The water category now has a 20 percent water reduction prerequisite. Previously, there was no water prerequisite.

- Four rating systems have been reweighted and now have 110 total points available, made up as 100 base points and 6 innovation and 4 regional points.

- In LEED V3-2009, there is an innovation credit for documenting sustainable cost impacts.

LEED for NC V3-2009

Since CSOs must operate in a LEED "pointchasing" environment, a broad understanding of the points and what they "buy" are important. The V3 version of the LEED NC Credit Categories are:

- Sustainable Sites   26 points
- Water Efficiency   10 points
- Energy & Atmosphere   35 points
- Materials & Resources   14 points
- Indoor Environmental Quality   15 points
- Innovation & Design Process   5 points
- LEED Accredited Professional   1 point
- Regional Priority   4 points

There are 110 points available. The levels of certification are:

Certified	40-49
Silver	50-59
Gold	60-79
Platinum	80 and above

SUMMARY REVIEW

A detailed description of prerequisites, credits and points can be obtained from the USGBC website as a pdf file at no charge. Better yet, for those very active in LEED projects, the USGBC offers an extremely thorough reference guide that can be purchased from the website. List price is $200 but pricing is discounted for member companies. The reference guide is a very detailed "How to do LEED" instruction manual and is highly recommended.

The LEED format for rating a green building consists of two categories:

- Prerequisites—These are mandatory requirements and *all must be satisfied before a building can be certified*!
- Credits—Each credit is optional, selected from a menu with each contributing to the overall total of points earned. This will determine the level a building will be rated; i.e. certified, silver, gold or platinum.

For a more focused look at each section, please refer to Appendix B.

In his article, "The LEEDing Way," Peter D'Antonio analyzes the activity in energy & atmosphere (E&A), and indoor environmental quality (IEQ), for the first 53 LEED Version 2.1 certified buildings:

- Regarding E&A, average points earned is only 5.3 out of the possible 17.
- This is the lowest percentage achieved in any of the five categories; so although E & A is the largest plum, few appear to be taking advantage of it.
- Renewable energy points are earned in fewer than 10 percent of the certified buildings.
- Regarding IEQ points, ventilation effectiveness and controllability are achieved in less than one third of buildings.

This article, which appeared in the May 2004 issue of *Energy User News,* provides support for the proposition that engineers are not maximizing their potential contributions to LEED buildings. Hence, there is a significant opportunity for the engineering profession to take on a larger role than we have had in the design, construction and operations of green buildings.

Other factors for the design team to consider are the forces that drive the LEED points on a project. Many times, for the design team, especially the mechanical, electrical, and plumbing (MEP) team members of a LEED project, it boils down to, "How many points can we get?" They control or influence approximately 75 percent of the total LEED credits on a job. Commonly called "point chasing," it is an effort by the design team to achieve the maximum available points at the minimum cost and effort. Although it is a rather ugly approach to green building design, it has become a fact. Teams will focus on points. Acquisition of points has become one of the elements in the design contract.

This approach can, and should, be managed by the CSO. It gets back to the integrated design process and the setting of goals during the design charette. During the charette process, the CSO should develop and prioritize the organization's goals for the project and the development of the points strategy to support those goals. For example, does the organization want very high energy efficiency, or does it want to make an environmental statement with a green roof?

Regarding "certified products," the term does not mean "LEED certified," but certified by other entities such as the Forrest Stewardship Council (FSC). Beware of manufactures claiming "LEED certified" products. The USGBC and LEED do not certify products. What they do is adopt industry standards as applicable, such as the FSC certified wood. If someone presents a product as "LEED Certified," they are being misleading. However, if a product is promoted as being LEED compliant, meaning that it that is capable of satisfying the particular requirements of LEED criteria, that is acceptable.

Why LEED-EB O&M Is So Important

Drawing on similarities to new construction program, LEED EB O&M has a larger potential impact and resultant benefits to society simply because there are many times more existing buildings than new construction.

Why the need for "New & Improved?" USGBC has reacted to marketplace desire to make LEED EB O&M more user friendly. Two of the hurdles that held back EB were the prerequisite for existing building commissioning and minimum energy performance as rated by EnergyStar Portfolio Manager. Owners have found that to perform these two operations can be costly. Further, if the results are poor; i.e., a low EnergyStar rating, it could be very expensive to upgrade the building's mechanical systems to achieve the required rating.

LEED 2009 EB O&M has the following minimum requirements:

- Comply with environmental laws;
- Be complete, permanent building or space;
- Use a reasonable site boundary;
- Minimum of 1,000 square feet;
- Serves more than 1 FTE occupant;
- Commit to sharing energy and water data; and
- Gross floor area to be no less than 2 percent of the gross land area within the LEED project boundary.

LEED EB: O&M has four levels of certification, which are compatible with the NC certification levels. They are:

- Certified Level   40 to 49 points
- Silver Level   50 to 59 points
- Gold Level   60 to 79 points
- Platinum Level   80 points and above

The point distribution for LEED-EB O & M are as follows:

	Points	
Sustainable sites	26	
Water efficiency	10	
Energy and atmosphere	35	
Materials and resources	14	
Indoor environmental quality	15	
	100	Total Points

In addition, points are given for:

Innovation and design process	4
Documenting sustainable building	
cost impacts	1
Regional priority	4
LEED accredited professional	1
Total Points Available	110

In summary, the LEED EB O&M is similar to LEED NC; however, all prerequisites must be satisfied, and the credits are optional depending upon the final points and certification level desired.

Energy Star Building Rating System and How It Fits with LEED EB O&M.

Why do we need LEED when we have EnergyStar? People are confused by the two. The answer is very simple, EnergyStar is exactly that, it is the energy efficiency of buildings. LEED encompasses a much broader spectrum of criteria for sustainable green buildings than just energy. Again this is a case of LEED not creating anything new, but simply adopting best management practices from the industry.

For ENERGY STAR rated building types:

• Use the ENERGY STAR Portfolio Manager benchmarking tool.
• Exceed the required ENERGY STAR rating of 69 points. Extra LEED points are shown below. This is a partial table, range is from 1 to 18 points, for scores of 71 to 95.

ENERGY STAR Score	LEED-EB Points
71	1
73	2
75 **	4
77	6
79	8
81	10
85	13
89	15
95	18

**Note: A score of 75 or more is required to earn the EnergyStar rating, again this is a separate rating but adopted by LEED as best management practices to determine the energy efficiency level of a building's performance..

THE LEED PROCESS

LEED is not simply a collection of parts, nor a collection of technologies. Maybe the most important aspect of LEED is not the points, but the holistic process that will deliver and maintain a high performance green and sustainable building.

• Design Team Integration
• Project Registration
• Project Certification
• Documentation

Critical to success is the integration of the design **TEAM** on Day 1. LEED is a marketplace transformer. It is a paradigm shift away from top down, minimum first cost emphasis. The hierarchical old-fashioned way was design, bid, build. The LEED design process is one of integrated, holistic building design, construction, operations and maintenance. All participate as equals on a construction roundtable with the following types of representation:

- Chief sustainability officer (CSO)
- Owner
- Architect
- Engineer
- Construction manager
- Contractors & subcontractors
- Equipment suppliers & manufacturers
- Commissioning authority—watchdog role

During the very early stages of a green building's development, a design charette should be held. This refers to meetings that are held over the course of a day or two, with follow-up meeting, wherein the entire team, the construction roundtable group, gets together to develop the roadmap to successful green buildings. The ENTIRE team joins in—all stakeholders, including the owner, CSO, designers, commissioning authority, and operations.

It is during the design charette, which occurs at the earliest moments of a project, that the energy engineer can provide the maximum overall benefit to the project. It is during this time that key choices are made about the lighting, HVAC, and building envelope. The energy engineer can help guide the team to the most appropriate energy efficient design strategies based upon the team's energy goals and the available energy sources.

MARKETING LEED AND SUSTAINABILITY TO THE COMMUNITY, OWNERS, AND DESIGNERS

It is vitally important that CSOs realize that LEED offers a great deal of value to a broad spectrum of the community. The owners benefit by having high performance buildings that are cost efficient and provide for better employee productivity. The design community benefits by having a way to craft a stronger value message for superior architecture and design by providing a way for designers to qualify and quantify their competitive advantage over other non-green designers. The community benefits by having a program that promotes urban and brownfield development, improves the environment, and provides for a healthy living style. Elaborating as to why a design team should be promoting LEED design, the overall life cycle cost of a typical commercial office building can provide a CSO with valuable insight.

The typical cost breakdown for a 40-year life cycle cost analysis is:

> Construction or First Cost is 11 percent
> Financing is 14 percent
> Alterations are 25 percent
> Operations are 50 percent

It is interesting that the cost most design teams grapple with, first cost, is actually the least significant cost element in the overall life cycle costs of a building. Thus, those decisions to keep first costs low by specifying cheaper designs and equipment can have a serious negative impact on the overall life cycle performance of a facility.

Additionally, consider that first cost is only 11 percent of the total life cycle cost and that A&E fees are only 6 to 8 percent of that 11 percent, or .88 percent of the total costs. These numbers suggest that it could be beneficial to the owner to pay more for superior architecture and engineering. This is because the designs and selections made by the design team have a great deal of leverage on the total life cycle ownership costs.

Impediments to Green Acceptance

Typically, the first and possibly most serious impediment to the wide-scale adoption of LEED is the perception that it costs more. The facts are that it may add cost, from 1 to 5 percent, depending upon the level of green the ownership team has identified in that design charette. But it does not necessarily cost more if the design team is clever about making design decisions and using all available resources that may be at hand.

Hint: If trying to promote LEED, look to identify market conditions in the project's locale that support LEED. Many states have

various programs to incentivize energy efficiency and other marketplace conditions which can affect the viability of a green project.
The following example from New Jersey shows how "market conditions" can drive LEED adoption.

CASE STUDY: LEED IN NEW JERSEY

First cost can be less of an issue if there are incentives for high
efficiency equipment. There is a program through the NJ Board of
Public Utilities called NJ Smart Start Buildings, which provides rebates for high efficiency equipment such as lighting, HVAC, boilers,
and chillers, as well as commissioning and design team meetings.
Essentially much of the cost differential between cheap inefficient
equipment and high efficiency equipment is offset by the rebates.

Renewable energy sources are promoted by statewide programs
such as the Clean Energy Program. Similar to Smart Start, renewables such as wind, solar PV, and biomass projects are rebated, some
up to 50 percent of the initial installed cost.

ASHRAE Std 90.1-2007 is the State Energy Code as well as the
prerequisite for Energy and Atmosphere LEED credits. So there is
no additional cost for NJ buildings to comply with this standard.
Although in other states that may not have this code requirement,
compliance with ASHRAE 90.1 might add cost.

Many brownfields are available for development with incentives
from the NJ Economic Development Administration (NJEDA). This
can be a simple prescriptive credit. NJ, as the most densely popu-
lated state in the nation, has many former industrial sites, which are
in inner city areas close to mass transit and part of urban renewal.
Thus, a brownfield site can facilitate a number of other credits.

Environmental—voluntary LEED adoption decreases need for
additional regulation. NJ is one of the most regulated and legislated
states, but with adoption of LEED, many of the goals of environmental legislation can be achieved voluntarily.

NJ has some of the highest energy costs in the nation; thus, payback periods on high efficiency equipment are shorter than other
states, which helps support the value of energy efficiency versus
low first cost equipment.

THE FUTURE OF LEED

After reviewing the potential for LEED in guiding sustainability, the CSO might well ask: Where is LEED headed? Or, where is the high performance green building industry headed? We can see there are new and revised rating systems and supporting standards coming out and/or in development.

There is a proliferation of new products and services centered on satisfying the demand for high performance buildings. Enhancements to LEED have been offered to accommodate marketplace demands, such as the incorporation of the optional Regional Priority Credits. There has also been more emphasis on energy efficiency, which helps to drive the economic benefits of LEED through energy cost savings.

Ten years ago, LEED was almost perceived as a cult. Certainly it was not mainstream. However, in less than a decade, there has been a revolution in how the design and construction of buildings are approached. Sustainability, green and LEED concepts have become well known in the industry. Even if a building does not go through the full LEED process, many of the sustainability concepts are considered in the process. LEED exerts such a strong influence in the buildings' industry that it has ratcheted up the level of sustainable design and performance for all buildings, not just those specifically planned to be green. Building on past successes, the sustainable and green concepts espoused in LEED are rapidly becoming mainstream.

Chapter 10

Sustainability Master Plan

James W. Brown and Shirley J. Hansen

If you were to lay this chapter and Chapter 10 from the book, *Investment Grade Energy Audit*,[10-1] side-by-side, you would see striking similarities. This is deliberate. We want to make the point that the move from an energy master plan to a sustainability master plan is smooth, relatively effortless, and very logical.

In the audit chapter, we called your attention to the need for sustainability auditors to carefully consider many related issues, including the implications for indoor air quality, water management, hazardous waste, etc. We have also declared that the sustainability procedures are inextricably linked with commissioning as well as measurement and verification. We have stressed that the auditor's clients should be provided with guidance regarding energy supply availability, energy security, and green supply chain criteria. Further, we have underscored the clear relationship of the audit to financing and, therefore, to implementation of a project.

These relationships suggest that any quality audit takes into consideration many facets of an operation and the sustainability audit itself takes on many roles and needs to reach into every corner of an enterprise. It is impossible to over-emphasize that, in comprehensive planning, measures cannot be taken in isolation.

As a sustainability auditor becomes more sophisticated, the report will increasingly lay the foundation for a master plan. It seems particularly fitting, therefore, that the last chapter of this book talks about the ingredients of a sustainability master plan and stresses how providing one allows the auditor to serve his or her client more effectively.

The best audit in the world does not save resources or improve the client's operation if it is not put to use. It seems an appropriate place to pull a quote from *Manual for Intelligent Energy Services*.[10-2]

> The audit is a valuable tool, but audits don't save energy, people do! The unattended audit report gathers dust. Only when it is read, discussed, and implemented can its energy/environmental/dollar benefits be realized. The difference between dust and energy savings is people. It is the communications connection that makes it work.

The kind of thinking put forth in this quote should become the conceptual backbone of any sustainability master plan (SMP). A plan's strategy should integrate the audit's technical findings into the tasks and responsibilities of the people who must implement the plan. The whole program should be held together by a strong communications component—spelled out in the plan.

Perennially, the weakest two components in an energy management program are people and communications considerations. Unfortunately, these weaknesses have spilled over into the sustainability program. For a client to get the maximum benefit of an audit, it is imperative that these two components be clearly spelled out. People and communications can make or break a plan that is *technically* perfect. As we have stressed throughout this book, the *management* aspect is critical. It is why we must focus on the chief sustainability officer (CSO) and his/her role. While we can move from an energy management plan to a sustainability master plan with considerable ease, we cannot turn an energy manager into a sustainability manager. The CSO job is broader and more demanding. Effective sustainability implementation requires changes in human behavior. This necessitates leadership and effective communication. The SMP becomes a critical tool in making it happen.

A good SMP, however, is much more than managing sustainability; it is managing sustainability *implications* throughout the facility and/or processes. We are also talking about managing the investment guidance the audit offers to assure that the maximum benefit of the sustainability audit is realized.

The sustainability management plan proposed in this book is a tool to be used in the development of an integrated plan for the intelligent purchase, operation, upkeep and replacement of resource consuming, and energy-related, equipment and systems. Although a SMP must be shaped to fit the needs of the owner or facility, a fundamental focus should include correction, renovation or replacement of particular resource consuming equipment/systems. These decisions must also weigh such concerns as safety codes/standards compliance, operating expenses, efficiency and equipment system age.

From the consultant's perspective, a master plan for the client is a value-added step beyond the audit. For the CSO it can be a game changer. If the audit identifies a weakness, the master plan explores ways to ameliorate or remove that weakness. Sometimes the plan can define ways to heal the breach, or augment the organization's resources.

Another plus for the consultant is the potential long-term relationship with a client that is engendered through the creation and implementation of a master plan. The cost of customer acquisition is always a factor in running a business. To the extent that a consultant can better serve his/her client while offering more services, the more profitable the relationship will be. And over time, op-

ENOUGH REPORTS TO SHED SOME LIGHT

portunities to sell even more services will present themselves.

In the global response to resource needs, it is imperative that we recognize that an SMP can, and should, focus on both energy efficiency and renewable energy. The "glamour" of renewables has tended to capture the spotlight and to pit renewable against energy efficiency. They are not, and should not be, at odds. They are compatible. The savings from energy efficiency can buy down the cost of renewables and make them a more economically viable option.

Even the most visionary auditor, CSO, or energy manager cannot dream up a plan while sitting at his, or her, desk. The full value of the plan's potential interrelationships comes from listening, looking, and gaining input from a broad range of people.

The auditor and those developing the SMP should view all facets of the client's operation through the lens of "resource availability." What are the vital resources used in a given process? How effectively are they being used? How can the usage be improved without having a detrimental impact on the process? The answers to these questions cannot be derived in isolation.

An auditor, who has effectively brought the people factor and risk considerations into his or her auditing procedures, is in an ideal position to guide the master plan development.

DEVELOPING THE SUSTAINABILITY MASTER PLAN

Let's first look at the typical ingredients of a SMP and identify those pieces that are "boiler plate." Then, we can weigh various factors that might help us individualize the plan to particular conditions. In many instances, it is like fitting together the pieces of a puzzle. Where do the audit, commissioning and M&V overlap, interlock, fit?

We fully recognize that not all SMPs are created for a single purpose, such as energy efficiency or wastewater treatment. Some plans are developed simply to form a process for planned replacement of major equipment and/or facilities. For example, a school district that finds itself losing enrollment and income may face the need to maximize the use of its building space without investing huge amounts of money in facility remodeling. The "master plan" for that district may center around questions such as:

- Which campuses do we close?
- Which campuses do we renovate?
- What are the steps toward renovation and how will the needed work be prioritized?
- What environmental consideration must be weighed in these decisions?

Other plans may be developed to create a pathway toward the accomplishment of a specific future goal, such as a hospital's decision to add an additional medical service requiring highly technical and expensive equipment as well as the skilled technical staff.

Far too often we find replacement recommendations being made to owners prescribing the exact opposite order of priorities. Let us give you a couple of examples: When money is made available to replace a portion of the building's equipment, many turn automatically to the oldest equipment simply because it "should be" the next item to fail. A piece of renewable equipment may be installed because government funds make it attractive—but the action will replace a perfectly good piece of equipment causing the whole process to work against the stated sustainability goals. Such decisions are frequently made because no one has gathered the data to really analyze the facts. Or, once gathered, have not taken the time to review the data from all perspectives. Which piece of equipment has caused the most trouble over the last few months? Which unit costs the most to maintain? Is there equipment that is oversized…undersized…and causes consistent comfort problems? Do we really need renewable equipment at this time?

An SMP should provide the information needed to make this decision based upon facts, not just gut-feel, and define a procedure for making those decisions. The bottom line is that the older boiler, operating within acceptable efficiency ranges and within acceptable maintenance expense ranges should be replaced *after* the newer low efficiency rooftop unit that consistently loses compression and never satisfies comfort requirements.

Or, we find equipment replacement budgets being spent for higher efficiency equipment just because it's higher efficiency equipment! The reasoning is sound. Increased efficiency produces cost savings that can, in turn, be used to replace other equipment. But, when energy efficiency projects take precedence over safety issues, the repercussions may far exceed the benefits of the efficiency investment.

Too often, CSOs are confronted with deferred projects that have been shelved because the cost to bring *ancillary systems* up to code were not included in the original project estimate? These are code/standard issues that simply cannot be ignored! Project cost overruns occur because the project cost estimator did not account for the safety and/or code compliance issues that *must* be done if any work is allowed at all.

When in the process of SMP development, the plan developer needs to keep in mind that, unless unlimited resources are available, budgeted funds must be spent in order of priority. Not the auditor's priorities...the owner's priorities! And the owner's priorities incorporate the *big picture*, not just the currently recommended project!

Most of the sustainability master planning discussion to this point has centered around the active systems that have to be dealt with and the power and other resources that make them work. To be truly effective, planning must go farther, and look at the structure within which all the active systems serve. To be complete, the plan must consider the needs of the building envelope. In fact, the building envelope will often determine the loads that the mechanical systems must satisfy. Further, the envelope components will determine how long the overall facility can remain in-place at an operational level.

In some cases, there is another consideration that can be a critical part of a master plan. The SMP should be a guiding light. Is there an intention at some future date to alter the usage of all or part of the structure, expand office areas, increase production space, add onto the building, or create major changes in heating/cooling loads? All of these things should be recognized in the plan. Some of them can have an impact on the part of the plan covering active systems. Prospective changes in the amount of office space or manufacturing space may alter HVAC replacement specification, the addition/ removal of power or water lines, or a dozen other things. To build a true SMP, we must remember that the structure holds all those things we treasure and is a vital part of the package.

A well-prepared SMP keeps the owner's priorities, life safety priorities and code/standard compliance priorities in focus at all times. It is the CSO's responsibility to see that the SMP blends *needed* projects so that they fit into the overall program right alongside the *wanted* projects.

Of all the benefits inherent within the sustainability master plan, possibly the most beneficial is *project prioritization*. The goal of this portion of the plan is to assess future expenses, planning them into future budgets and reducing unexpected costs and budget-busting emergencies. Although no one can flawlessly predict the order of demise for individual equipment items, it has been found that, given data such as age, degree of maintenance, hours of operation, etc., one's batting average can be significantly improved.

GUIDING LIGHT

ADDED SMP COMPONENTS

Additional components that may be incorporated into the sustainability master plan are:

Engineering Survey
An analysis of general system types, installations, problems and renovations needed for:

- Mechanical Systems: Heating, cooling, ventilation, refrigerant type, air and water distribution, domestic hot water, etc.

- Electrical Systems: From the transformer into the building, down to the power available in each building area. Electrical systems, like mechanical, wear out over time...they become overloaded as building loads increase...they should be analyzed and upgraded, even if no energy savings are to be obtained.

- Plumbing Systems: The type of fixtures, need for hot water, water conservation, ADA Compliance, odor sources—all need to be a part of the SMP.

- Structural Systems: Building envelope, including doors, windows, siding, roof, etc.

- Information Technology Systems

- American Disabilities Act (ADA) Issues

- Indoor Air Quality Issues

- Water Management: As water meters get older, measurement typically favors the supplier.

- Wastewater Treatment

- Renewable Options

- Environmental Impact of Certain Options

A key goal in achieving true sustainability for your equipment is to classify the types of systems and estimate, as closely as possible, the remaining useful life of the equipment. During this task, an analysis of overall systems and the inter-relationship between systems is conducted.

Inventory of Systems

A specific analysis of individual equipment items should be gathered including:

- Nameplate data: Serial number, electrical information, motor size ... know what you have!

- Area served: know which system is causing the problem.

- Equipment age: estimate the length of life remaining.

- Equipment condition: what can be anticipated from this unit?

- Equipment maintenance history: what does it *really* cost to operate this unit?

- Suitability of the application for that equipment: is there something newer...better?

The SMP developer should dig out as much equipment history as possible. Work orders, maintenance records, operating expenses, frequency of problems, types of problems—all this information is needed when preparing recommendations for renovation programs. With this information, several of the questions raised while analyzing the overall systems can be answered. Why particular control strategies aren't working could be explained by the discovery of inoperable control valves or disconnected economizer damper linkages. This research can prove to be exceedingly valuable to long range planning.

Another obvious benefit to this step is the accumulation of data describing the equipment being used. An inventory such as this can prove to be very beneficial when attempting to decide which equipment replacement should be inserted into the budget two, three or even four years down the road.

Energy Analysis

This analysis serves to help the CSO know precisely where his/her energy is being used and how much is consumed by specific items of equipment. When the continuous presence of power is an absolute necessity, an analysis of the emergency generation equipment is needed. In addition, a study of the various types of fuel sources available, their dependability, and the potential for providing the same service through multiple fuel options becomes a high priority task. An Indian minister of power once offered us some food for thought, "No power is as costly as no power." Alongside the security of the energy supply is its cost; however, when power is absolutely essential to an operation, availability will weigh more heavily than cost. Electric deregulation, utility contract negotiations, or simply explanations of the way you are being billed for utilities can be included in this portion of a SMP.

Water Analysis

An analysis of where water is being used, the facilities that are using that water, etc. needs to be done. Consumption should be compared to benchmark data as underground pipe leakage is surprisingly common. Equipment exists to check underground piping without disturbing the ground.

Wastewater Treatment

A surprising amount of energy is devoted, some of it unnecessarily, to treating wastewater. For example, wastewater treatment plant supervisors are inclined to set the aeration system to meet the EPA standards for worst bio-effluent conditions because it is a "safe" way to satisfy regulators. Software exists that reads the diodes and can adjust the aeration accordingly—similar to an energy management system is a building. Long ago, software that cost only $75,000 saved the Northwest Memphis Wastewater Treatment Plant $750,000 in the first year. Imagine the reduction in energy consumption and pollution emissions, too. Yet, the reluctance to install such software persists.

Hazardous Waste

Every CSO must face the need to reduce waste economically and effectively. This is particularly true for hazardous waste. Under a performance contract with Johnson Controls, part of the project was installing an incinerator which turned hazardous waste into general waste which saved a hospital over $1 million per year.

Operations & Maintenance Evaluations

O&M evaluation plans should include:

- Development of maintenance schedules for the equipment located during the inventory;

- Preparation of man-hour estimates required to provide appropriate levels of maintenance;

- Preparation of labor and parts costs for maintenance using recent parts billings, local area price proposals, owner supplied hourly rates and overhead costs;

- Creation of, or revision to, the maintenance department's work order system; and

- Preparation of a preventive maintenance program for primary equipment items to improve operation and lengthen equipment life spans.

Development of Software

Master planning software should be designed to include an inventory of individual equipment information including maintenance costs, parts numbers, operation recommendations, etc.

This software should provide ease of access and be integrated into the work order system to track frequency of work provided and the degradation of equipment efficiency over time. It is also very beneficial during budget preparation to determine the specific costs of equipment scheduled for replacement in the coming year.

Procurement Standardization

The number of systems that must be learned by the maintenance department and the amount of inventory that must be stored can be reduced significantly by standardizing the types of equipment and preferred components. At the same time, this standardization can help insure consistent efficiency levels in new purchases. Large manufacturers increasingly depend upon suppliers to store repair and replacement parts.

It is important to insist that the supply chain document sustainable practices. Criteria should be developed and used. Wal-Mart's efforts in this regard serve as a good model.

Measurement & Verification

Broadly accepted M&V protocols are needed for effectively communicating program results. CSOs will not know what is working effectively, what the trends are, how good the engineering and O&M efforts actually are, or how much smaller its carbon footprint has become without measuring and verifying its procedures,

Training of Staff

Staff, and to some extent, occupants, need information regarding proper system operation, policy limits on temperature ranges and humidity levels. A re-evaluation of operating hours for various kinds of equipment during various times of the day and year need periodic assessment.

Communications Component

We have underscored throughout this chapter that a good audit offers a framework for much of the SMP. This is especially true if we take into account the "People Factor"[10-3] in a quality investment grade audit. A truly effective sustainability program must have a strong communications component and the SMP should outline the strategies to make it work.

COMMUNICATION

COMMON COMPONENTS OF AN SMP

In broad terms, a sustainability master plan should include:

- An assessment of current resource use and its implications for operation/mission;
- An inventory of the energy and water consuming equipment and systems;
- Existing and code operating parameters; e.g., air changes per hour (or cfm/occupant), temperature ranges, humidity levels, lighting levels, etc.;
- Baseline information on what current consumption is and what conditions, such as occupancy, run times, operating hours, cause that consumption;
- An initial indication of efficiency and/or savings potential (scoping audit) and specific organizational benefits (energy, water, waste) which could result from such savings;
- Environmental concerns: indoor air quality issues; pollution and emission considerations;
- An assessment of supply options, including pricing, trends, and availability;
- Emergency preparedness and standby provisions;

- An analysis of the operations and maintenance function; manpower, skills, training needs and related energy implications;
- Equipment replacement recommendations (including priorities by year);
- Criteria for green supply chain;
- A communications component;
- A determination of what needs can best be served by in-house staff and what tasks will need to be outsourced; and
- A sustainability policy.

Always a Work in Progress

A SMP should be a dynamic document ready to meet changes to operations or external forces which may impact it. It will be most effective when it is flexible and responsive to potential changing conditions. At a minimum, the SMP should be reviewed, and revised as appropriate, on an annual basis.

Whether the facilities in question are schools, hospital, industrial parks, factories, or state facilities, buildings are a collection of brick, limestone, wooden boards, and glass held together with glue, putty, mortar and nails. Sometimes they hold processes where widgets are combined with belts and motors to produce gizmos. No matter what labels are put on the edifices, products, or functions, the master plan needs to recognize that all the buildings and processes are part of an investment portfolio, which the CSO manages. The audit identifies ways to increase the value of this portfolio. The master plan provides guidance in how to actually implement these recommendations as effectively as possible. With a long-term view, such as a five-year plan, the document will also address ways to maintain and enhance these investments.

Any SMP will always be viewed by the various people in the organization through the perspective of their position in the organization and their responsibilities. The development of an effective plan will recognize the range of perspectives and the need for the affected people to understand what is being suggested. For example, recognition of a plan to enhance the organization's investment

"SOMEDAY THIS WILL ALL BE SUSTAINABLE"

portfolio will appeal to top management and the financial people. It will do less for the facility managers and plant engineers, who are charged with developing and maintaining productive work environments.

Whether we are talking about running motors, turning on lights, fixing dripping faucets, conditioning air, or using energy as a raw material, intelligent use of resources requires management. Effective management requires knowledge and skills that seem to grow exponentially. As they grow, they require even greater support, authority and budget to make things happen. An absolutely essential part of any plan, therefore, is a statement of the resources, CSO's responsibilities, the associated authority and budget. As a result, we conclude this chapter with comments regarding the importance of an enforceable policy, designed to illustrate senior management's commitment to the program.

THE ALL IMPORTANT SUSTAINABILITY POLICY

Although much has been written about sustainability and its importance to the local and global environmental needs, it needs to be stated that, in our opinion, *at least one more time,* that nothing has more impact upon the success, or failure, of a sustainability program than this policy. If prepared correctly, the sustainability policy will ensure that the audit report is used effectively and the sustainability master plan will begin, *and continue to be,* as a credible influence upon the facility and its occupants.

The Sustainability Policy's basic components are:

- Goal Statement: Specific, challenging yet attainable goals stated with the authority of the governing management.

- Plan Statement: The method(s) selected by administration to accomplish the goals, including responsibilities, authority and budget for those implementing the policy.

- Specific Issues: A comprehensive list of major issues that are to be resolved.

- An indication as to how success will be measured.

- Reporting: Frequency and content of reports required to ensure success of the program.

Only after the distribution of this administrative declaration can anyone be held responsible for producing the desired results. If given no authority or insufficient budget to implement the program, the CSO cannot be held responsible for the lack of results.

The real issue faced by members of senior management when considering sustainability is commitment. Are they committed to the success of the program, and are they willing to impart the authority and allocate the funds necessary to assure the success of the program? Only through the publication of such a commitment, *by the people who can really enforce it,* can we ever expect to create a truly effective program.

Appendix A

Contributors

The authors are exceedingly aware of the value our contributors have made to this book, and wish to acknowledge them with the brief bios presented below. They are presented in alphabetical order. Each bio is followed by an email address should the reader wish to contact them directly.

BOB DIXON

Bob is an industry veteran and visionary, with over 34 years of experience in building systems, facility operations, energy conservation and management.

Currently serving as Senior Vice President and Global Head of Efficiency & Sustainability at Siemens Industry, Inc., Building Technologies Division, Bob leads all strategic global initiatives on building efficiency and sustainability including strategic planning, operations, product/solution development, program implementation, business development, and mergers and acquisitions. Bob is the first and currently only designated Senior Principle Expert for the 39,000 employee Building Technologies division.

Bob is a past president of the National Association of Energy Service Companies (NAESCO), and currently serves as Industry First Vice Chair for the Alliance to Save Energy. He is the Industry Member on the Buildings and Appliances Task Force under the Asia-Pacific Partnership on Clean Development and Climate, and leads Building Technologies Division's strategic global partnership with the Clinton Climate Initiatives. A well respected and sought-after speaker on energy efficiency topics around the globe, Bob earned his Bachelor of Science degree in Mechanical Engineering from California Polytechnic State University in San Luis Obispo, California, and is a graduate of the Minnesota Executive Program at the University of Minnesota.

bob.dixon@siemens.com

STEPHEN HANSEN

A Pacific Northwest native, Stephen Hansen has been working as a professional artist since 1968. While best known for his paper-mâché sculpture, he also works in wood, bronze, steel, stone, and resin among other materials.

Hansen has had one-man shows in commercial galleries and art museums in Chicago, Detroit, Los Angles, New York, San Francisco, Santa Fe, Washington DC, Naples, Caracas, and Sri Lanka. His work is included in many corporate, government (including the Smithsonian), and private collections in the U.S., Europe, South America and Asia. In 2009 his work was added to permanent fine art collection of The Federal Reserve Board in Washington DC, and he received the New Mexico "2009 Governor's Award for Excellence in the Arts."

He is professionally committed to the notion that art can be serious without being somber.

hansenart@comcast.net

STEPHEN A. ROOSA

Dr. Roosa is presently an Account Executive with Energy Systems Group, a full-service energy savings performance contracting (ESCO) company with offices in Louisville, Kentucky. His past experience includes energy savings assessments for over 3,500 buildings. His work has included energy efficiency, energy conservation and renewable energy projects. Considered an expert is sustainability, he teaches seminars throughout the U.S. in Sustainable Development and Renewable Energy.

He is a past President of the Association of Energy Engineers (AEE), associate member of the AIA, and the author of *The Sustainable Development Handbook*, a book on sustainability and sustainable design practices. He is a LEED Accredited Professional, a Certified Sustainable Development Professional, and a Certified Energy Manager. He holds a Doctorate in Planning and Urban Development, an MBA and a Bachelor of Architecture degree.

StephenRoosa@cs.com

ROBERT J. SANSONE

Mr. Sansone is a recognized international and domestic expert in the areas of risk identification, assessment and modification. His insurance experience includes heading up the global risk engineering organization of XL Insurance and the VP of Energy Engineering at The Hartford Steam Boiler Inspection and Insurance Company.

In 2008, he founded The Power Gen and Construction Practice, LLC to specifically address the need for thorough and proper risk assessment in these industries. His experience on the operations side includes senior level positions within The Adolph Coors Co., Air Products and Chemicals, Inc, United Engineers and Constructors, Inc. and the American Ref-Fuel Co. (a firm that designed, built, owned and operated waste to energy facilities).

He is a 1973 graduate of the United States Military Academy at West Point, where he earned a Bachelor of Science degree.

rjsenergy@gmail.com

NICK STECKY

Nick Stecky is president of NJS Associates LLC. His firm focuses on energy, sustainability, and green, high performance facilities. He received his BS in Engineering from the NJ Institute of Technology and a MS Systems Science from Fairleigh Dickinson University. He holds accreditations as a Certified Energy Manager (CEM), Leadership in Energy and Environmental Design (LEED) and a Six Sigma Green Belt. He has over thirty years of experience in the construction, operation and management of various facilities including industrial, process, corporate headquarters and R & D Facilities. The focus of these activities has been primarily on energy, environment, cost reduction and productivity gains. Nick currently serves as the Resource Efficiency Manager at Picatinny Arsenal for the Department of Defense providing Sustainable Design and Development services. Other experiences include Johnson Controls, General Electric, T.J. Lipton, JCP&L, PSE&G, and Princeton University's Plasma Physics Lab.

Leadership activities include being one of the founding members of the NJ Chapter of the US Green Building Council. He served two years as the Chair of the ASHRAE Technical Resource Group 7, TRG-7, for Sustainable Buildings and Operations. He is a past president of the NJ Chapters of ASHRAE and the IAEE. He was elected 2006 Region 1 Vice-President for the IAEE, and named IAEE Regional Energy Manager of the Year in 2006. Recently, he was recognized with the IAEE's 2010 Energy Professional Development Award.

nstecky@aol.com

Appendix B

Sample LEED
Assignment of Points

LEED NC SUSTAINABLE SITES
(26 Possible Points)

Prerequisite: Construction Activity Pollution Prevention
Credit Categories:
* Site selection—1 point
* Development density & community connectivity—5 points
* Brownfield redevelopment—1 point
* Alternative transportation (includes public transportation access, bicycle storage, changing rooms, low emission and fuel efficient vehicles, parking capacity)—up to 12 points
* Site development—protect or restore habitat, maximize open space—up to 2 points
* Stormwater design—quality and quantity control—up to 2 points
* Heat island effect—roof and non-roof applications—up to 2 points
* Light pollution reduction—1 point

WATER EFFICIENCY
(10 possible points)

Prerequisite: Water Use Reduction 20 percent
Credits:
* Water efficient landscaping—50 percent reduction, no irrigation, non-potable water irrigation—2 to 4 points
* Innovative wastewater technologies—2 points
* Water use reduction—reductions of 30, 35, or 40 percent earn 2, 3 or 4 points respectively

ENERGY & ATMOSPHERE
(35 possible points)

Three Prerequisites:
Fundamental Commissioning Building Energy Systems
Establish Minimum Energy Performance, 3 options for compliance:
- Whole building simulation—demonstrate a 10 percent reduction in NC or a 5 percent reduction in major renovations below the baseline energy consumption developed in accordance with the ASHRAE Standard 90.1-2007
- Prescriptive compliance path: Comply with all applicable criteria of ASH RAE Advanced Energy Design Guide, AEDG. Must use the particular AEDG for the application, such as AEDG for small office, and conform to the eligibility requirements. Other guides exist for schools, retail, and warehouses.
- Comply with the prescriptive compliance path of the Advanced Building Gore Performance Guide.

Credits:
- Optimize energy efficiency—possible 19 points
- On-site renewable energy—7 points possible depending on the percentage of total energy cost, from 1 to 13 percent
- Enhanced commissioning—goes beyond the commissioning required by the prerequisite, this is whole building commissioning which is initiated at the very beginning of a project—2 points
- Enhanced refrigerant management—reduce or avoid the use of ozone depleting compounds such as HCFCs—2 points
- Measurement & verification plan—3 points
- Green power—a 2-year renewable power purchase contract for at least 35 percent of the building's electricity—2 points

MATERIALS & RESOURCES
(14 possible points)

Prerequisite: Storage & Collection of Recyclables
Credits:
- Building reuse—3 points possible at 55 or 95 percent reuse of Existing walls, floors and roof.

- Building reuse, retention of 50 percent interior non-structural elements, 1 point.
- Construction waste management—2 possible points at diverting 50 or 75 percent
- Material reuse—2 possible points at 5 or 10 percent
- Recycled content—2 possible points at 10 or 20 percent
- Local/regional materials—2 possible points if manufactured and extracted, harvested, or recovered within 500 miles; or harvested locally
- Rapidly renewable materials—1 point
- Certified wood—1

INDOOR ENVIRONMENTAL QUALITY
15 possible points: no change from v2.2

Prerequisite 1—Minimum IAQ Performance—compliance with ASHRAE Standard 62.1-2007
Prerequisite 2—Environmental Tobacco Smoke Control, ETS
Credits:
- Outdoor air delivery monitoring—1 point
- Increased ventilation—1 point
- Construction IAQ management plan—2 possible points, during construction and before occupancy
- Low emitting materials—4 possible points relating to adhesives, sealants, paints, carpet, composite wood
- Indoor chemical & pollutant source control—1 point
- Controllability of systems—lighting, thermal comfort—2 points available
- Thermal comfort—comply with ASHRAE Std 55-2004—1 point
- Thermal comfort—verification—1 point
- Daylighting & views—daylight for 75 percent of spaces—1 point
- Daylighting & views—views for 90 percent of spaces—1 point

INNOVATION & DESIGN PROCESS
6 possible points

- Use of LEED-accredited professional-one point
- Innovation in design—five possible points. The intent of these

Innovation Credits is to provide design teams and projects the opportunity to be awarded points for exceptional performance above the requirements set by the LEED-NC Green Building Rating System. The USGBC, in particular, encourages the submission of innovative ideas in energy reduction and water savings.

REGIONAL PRIORITY
4 possible points

V3-2009 recognizes and encourages regionalization innovation, called "Regional Priorities," There are 6 Regional Priorities, but only a total of 4 can be selected to earn a maximum of 4 points. These include:
- Innovation in design
- Indoor environmental quality
- Materials & resources
- Energy & atmosphere
- Water efficiency
- Sustainable sites

The following were identified by the USGBC's regional councils and local chapters to address geographically specific environmental priorities:
- Credits Identified by the USGBC Regional Councils as having importance for a project's region. Determined by Zip code.
- Database of regional credits is available on the USGBC website; www.usgbc.org.
- Examples include credits for water conservation in arid zones, preservation of prime agricultural lands, reuse of existing building stock.

References

Chapter 1

1. GreenBiz.com 2/1/2010
2. http://www.greenbiz.com/blog/20 10108/02/introducing-ule-880 sustainability—manufacturing

Chapter 3

1. Unsworth, Malcolm. "Itron's Point of View" *EnergyBiz* magazine, May/June 2010. P. 44.
2. David W. Crane, President and CEO, NRG Energy, Inc. Testimony before the US Senate Committee on Environment and Public Works, October 28, 2009.
3. evo-world.org
4. "Corporate Water Accounting: An analysis of Methods and Tools for Measuring Water Use and Its Impacts" by Jason Morrison *et al.*
5. Allan, Sharon and Susan Christensen, "the utility's smart grid journey:; an odyssey of pervasive change," *EnergyBiz*, Volume 7, Issue 3. May/June 2010. p. 48.
6. *Ibid.*
7. Lloyd's *Global Water Scarcity: Risks and Challenges for Business.* April 2010.
8. Barton, Brooke. *Murky Waters? Corporate Reporting on Water Risk.* Ceres 2010.
9. Pullen, Jean. "City of Atlanta, RM Clayton WWTP Combined Heat & Power System Business Case." Unpublished. July 2010.
10. Celebrity Cruises, *2008 Stewardship Report.*
11. AAA, Journey. March/April 2010 p. 14 www.AAAjourney.com
12. Cafazzo, Debbie. TNT 5/18/20101 debbiecafrazzo@thenewstribune.com
13. Energy Business Reports. http://webmail1.webmail.aol.com/31650-111/aol-1/en-us/mail/print message.asph

Chapter 4
1. Boyd, Tonya. "Renewable to the Core: OIT taps campus geothermal resources" *District Energy. First Quarter 2010,* A publication of the International District Energy Association.
2. Energy Business reports, "Renewable Energy Government Incentive Programs." 2/21/2010

Chapter 5
1. U.S. Green Business Council October 2006
2. Bowman, A. (2003, August 28-31). From Green politics in the city. Presentation at the annual meeting of the American Political Science Association. Philadelphia.
3. Governor's Green Government Council. What is a green building? Building Green in Pennsylvania. http://www.gggc.state.pa.us/gggc/lib/gggc/documents/whatis041202.pdf, accessed 18 January 2007.
4. *Ibid.*
5. Siemens AG. (2007). Combat climate change-less is more. http://wap.siemensmobile.com/en/journal/story2.html, accessed 3 July 2007.
6. The average outside air temperature during the study period was 46 degrees F. The assessment was based on billings provided by the local utility. It is likely that the substantially larger electrical usage experienced by the newer building was also due, in part, to two desktop computers which operated almost continuously during the study month, and to the larger area of interior lighting. The newer brick building also had 11 outside light fixtures while the older frame building only had 2.
7. USEPA. Superior energy management creates environmental leaders, Guidelines for energy management overview. 5www.energystar.gov/index.cfm?c=guidelines.guidelines_index, accessed 3 January 2007.
8. USEPA. Portfolio manager overview. January 2007 www.energystar.gov/index.cfm?c=evaluate_performance.bus_portfolimanager, accessed 3 January 2007.
9. Lincoln Hall, a LEED building at Berea College in Berea, Kentucky, actually uses copper downspouts.
10. For a listing and explanation of the various rating systems developed for LEED, see the U.S. Green Building Council

rating systems website. www.usgbc.org/Display/Page. aspx?CMSPageID=222, accessed 2 January 2007.

11. A new 2007 edition of the IPMVP has been developed by the Efficiency Valuation Organization (EVO). See http://www.evo-world.org/index.php?option=com_content&task=view&id=60&ltemid=79.

12. Kentucky's first LEED-NC school, seeking a Silver rating, was initially estimated to cost over \$200/ft^2 (\$2,152/m^2) compared to the local standard costs of roughly \$ 140/ft^2 (\$1,505/m^2) for non-LEED construction.

13. International Code Council (2003). International Energy Conservation Code. p. iii.

Chapter 9

1. ANSIIASHRAEIUSGBC/IES Standard 189.1-2009 Standard for the Design of High-Performance Green Buildings

2. ASHRAE GreenGuide, 2003—Editor David L. Grumman

3. ASHRAE Advanced Energy Design Guide

4. ASHRAE Website www.ASHRAE.org

5. *Energy User News,* May 2004 by Peter D'Antonio, entitled "The LEEDing Way."

6. Fowler, KM and Rauch, EM, 2006, Sustainable Building Rating Systems: Summary—PNNL-15858, Pacific Northwest National Lab.

7. The United States Building Council, www.USGBC.org

Chapter 10

1. Hansen, Shirley J. and James W. Brown. *Investment Grade Energy Audits: Making Smart Energy Choices.* Published by The Fairmont Press, Lilburn, Georgia. 2004.

2. Hansen, Shirley J. *Manual for Intelligent Energy Services.* Published by The Fairmont Press, Lilburn, Georgia. 2002, pages 27-28.

3. Hansen, Shirley J. and James W. Brown, *op. cit.* page 41

Index